Ph. Tondeur

Foliations on Riemannian Manifolds

Springer-Verlag

Universitext

Editors

J. Ewing
F.W. Gehring
P.R. Halmos

Universitext

Editors: J. Ewing, F.W. Gehring, and P.R. Halmos

Booss/Bleecker: Topology and Analysis
Charlap: Bieberbach Groups and Flat Manifolds
Chern: Complex Manifolds Without Potential Theory
Chorin/Marsden: A Mathematical Introduction to Fluid Mechanics
Cohn: A Classical Invitation to Algebraic Numbers and Class Fields
Curtis: Matrix Groups, 2nd ed.
van Dalen: Logic and Structure
Devlin: Fundamentals of Contemporary Set Theory
Edwards: A Formal Background to Mathematics I a/b
Edwards: A Formal Background to Mathematics II a/b
Endler: Valuation Theory
Frauenthal: Mathematical Modeling in Epidemiology
Gardiner: A First Course in Group Theory
Godbillon: Dynamical Systems on Surfaces
Greub: Multilinear Algebra
Hermes: Introduction to Mathematical Logic
Humi/Miller: Second Order Ordinary Differential Equations
Hurwitz/Kritikos: Lectures on Number Theory
Kelly/Matthews: The Non-Euclidean, The Hyperbolic Plane
Kostrikin: Introduction to Algebra
Luecking/Rubel: Complex Analysis: A Functional Analysis Approach
Lu: Singularity Theory and an Introduction to Catastrophe Theory
Marcus: Number Fields
McCarthy: Introduction to Arithmetical Functions
Mines/Richman/Ruitenburg: A Course in Constructive Algebra
Meyer: Essential Mathematics for Applied Fields
Moise: Introductory Problem Course in Analysis and Topology
Øksendal: Stochastic Differential Equations
Porter/Woods: Extensions of Hausdorff Spaces
Rees: Notes on Geometry
Reisel: Elementary Theory of Metric Spaces
Rey: Introduction to Robust and Quasi-Robust Statistical Methods
Rickart: Natural Function Algebras
Smith: Power Series From a Computational Point of View
Smoryński: Self-Reference and Modal Logic
Stanisić: The Mathematical Theory of Turbulence
Stroock: An Introduction to the Theory of Large Deviations
Sunder: An Invitation to von Neumann Algebras
Tolle: Optimization Methods
Tondeur: Foliations on Riemannian Manifolds

Philippe Tondeur

Foliations on Riemannian Manifolds

Springer-Verlag
New York Berlin Heidelberg London Paris Tokyo

Philippe Tondeur
Department of Mathematics
University of Illinois
Urbana, IL 61801, USA

With 7 Illustrations.

Mathematics Subject Classification (1980): 53C12

Library of Congress Cataloging-in-Publication Data
Tondeur, Philippe.
  Foliations on Riemannian manifolds.
  (Universitext)
  Bibliography: p.
  Includes indexes.
  1. Foliations (Mathematics) 2. Riemannian
manifolds. I. Title.
QA613.62.T64 1988   516.3′6   88-2011

© 1988 by Springer-Verlag New York Inc.
All rights reserved. This work may not be translated or copied in whole or in part without the written permission of the publisher (Springer-Verlag, 175 Fifth Avenue, New York, NY 10010, USA), except for brief excerpts in connection with reviews or scholarly analysis. Use in connection with any form of information storage and retrieval, electronic adaptation, computer software, or by similar or dissimilar methodology now known or hereafter developed is forbidden.
The use of general descriptive names, trade names, trademarks, etc. in this publication, even if the former are not especially identified, is not to be taken as a sign that such names, as understood by the Trade Marks and Merchandise Marks Act, may accordingly be used freely by anyone.

Camera-ready text prepared by the author.
Printed and bound by R.R. Donnelley & Sons, Harrisonburg, Virginia.
Printed in the United States of America.

9 8 7 6 5 4 3 2 1

ISBN 0-387-96707-9 Springer-Verlag New York Berlin Heidelberg
ISBN 3-540-96707-9 Springer-Verlag Berlin Heidelberg New York

QA
613.62
.T64
1988

To Claire

PREFACE

A first approximation to the idea of a foliation is a dynamical system, and the resulting decomposition of a domain by its trajectories. This is an idea that dates back to the beginning of the theory of differential equations, i.e. the seventeenth century. Towards the end of the nineteenth century, Poincaré developed methods for the study of global, qualitative properties of solutions of dynamical systems in situations where explicit solution methods had failed. He discovered that the study of the geometry of the space of trajectories of a dynamical system reveals complex phenomena. He emphasized the qualitative nature of these phenomena, thereby giving strong impetus to topological methods.

A second approximation is the idea of a foliation as a decomposition of a manifold into submanifolds, all being of the same dimension. Here the presence of singular submanifolds, corresponding to the singularities in the case of a dynamical system, is excluded. This is the case we treat in this text, but it is by no means a comprehensive analysis. On the contrary, many situations in mathematical physics most definitely require singular foliations for a proper modeling. The global study of foliations in the spirit of Poincaré was begun only in the 1940's, by Ehresmann and Reeb.

What is done in these notes is first to introduce foliations in Chapters 1 through 4 and then, in Chapters 5 through 13, to explore some interactions of foliations with the Riemannian geometry of the ambient manifold. The Riemannian foliations, introduced by Reinhart in 1959, are of particular interest.

Chapter 1 contains motivations for the theory of foliations, and ends with a list of books and surveys on particular aspects of foliations on pages 5-6. Chapter 2 is a discussion of an instructive special case, namely transversally oriented foliations of codimension one. The precise definition of a foliation follows these examples and appears at the beginning of Chapter 3. In Chapter 4 some ideas centering around the concept of holonomy are sketched. The infinitesimal aspects of this are encoded by the connection in the normal bundle defined in Chapter 5 and used throughout the following chapters. From a technical point of view this connection contains all the information from Chapter 4 subsequently used. The prerequisites for the first four chapters are the calculus of differential forms on smooth manifolds (see pages 22-23 for a list of formulas and conventions used), the integrability theorem of Frobenius, and the theorem of De Rham identifying the cohomology of differential forms with singular cohomology with real coefficients.

A reader familiar with the rudiments of foliation theory may want to begin directly with Chapter 5. The prerequisites at this stage include the concepts of connections in vector bundles and curvature, and elementary topics of Riemannian geometry. There are frequent references to basic facts of algebraic topology. The last four chapters presuppose some familiarity with Lie groups and their Lie algebras. Chapter 12 contains a generalization of the classical De Rham-Hodge decomposition theorem for differential forms. The proof, as well as the material in Chapter 13, requires some familiarity with ideas from the theory of elliptic partial differential equations. The list of references to topics touched upon in this text is followed by a bibliography on the general subject of foliations (a large list, but certainly incomplete).

There is a fast growing literature on the subject of Riemannian foliations, and these notes do not, by any means, cover all aspects of even this limited field. They represent the text of lectures given at the University of Illinois during the fall of 1986, and they were written as the lectures progressed. It is a pleasure to thank Hilda Britt, Cherri Davison, Lori Dick and T^3 for their care in typing this text. Thanks are due to Byoung Keum, Seiki Nishikawa, Paul Scofield and the referee for the elimination of errors. I also take this opportunity to express my gratitude to the University of Illinois at Champaign-Urbana and the National Science Foundation for their continued support.

Philippe Tondeur

INTERDEPENDENCE OF CHAPTERS

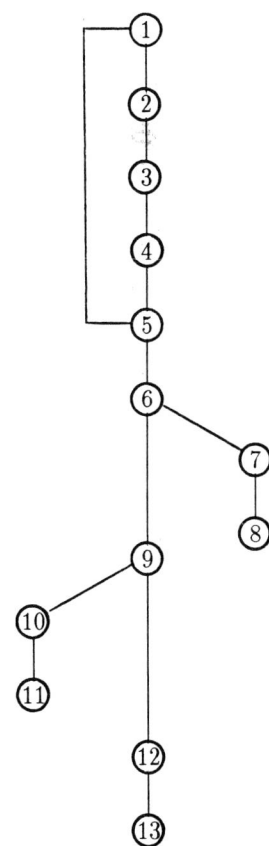

CONTENTS

Preface..vii
1. Introduction...1
2. Integrable forms...8
3. Foliations..24
4. Flat bundles and holonomy...35
5. Riemannian and totally geodesic foliations..............................47
6. Second fundamental form and mean curvature..............................62
7. Codimension one foliations..74
8. Foliations by level hypersurfaces......................................104
9. Infinitesimal automorphisms and basic forms............................117
10. Flows..132
11. Lie foliations..143
12. Twisted duality...149
13. A comparison theorem..164
References..169
Appendix: Bibliography on foliations......................................179
Subject index...243
Index of notations..246

CHAPTER 1

INTRODUCTION

One way to think of a foliation is to think of it as a higher dimensional dynamical system. A dynamical system or vector field X on a smooth manifold M gives rise to a decomposition of M by the integral curves of X. Outside the singular set of X there is precisely one maximal connected 1-dimensional integral curve of X passing through each point.

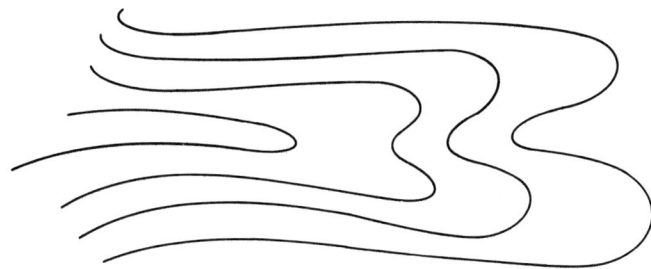

Figure 1.1

A (nonsingular) foliation $\mathcal{F}$ on a manifold M can approximatively be thought of as a partition of M into p-dimensional submanifolds of M (see p. 24 for a complete definition). The submanifolds are the leaves of the foliation. They are not meant to be necessarily embedded in M. E.g. the Kronecker line on the 2-dimensional torus T^2 gives an example of a 1-dimensional foliation where every leaf is dense in T^2. A good example to keep in mind is the case of a foliation by hypersurfaces, that is to say a foliation of codimension one (this situation is discussed in Chapter 8). In general, the codimension of a foliation is denoted by q, where $q = n-p$. The two extreme cases $p = 1$ and $q = 1$ are the simplest and best understood cases.

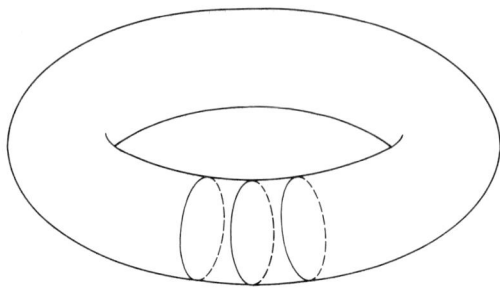

A foliation of the torus T^2 by curves (p = 1).

Figure 1.2

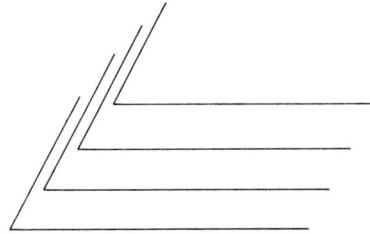

A foliation of $\mathbb{R}^3$ by parallel planes (q = 1).

Figure 1.3

Another way to view a foliation is as a system of first order partial differential equations (ordinary DE for p = 1), the solutions of which are then the leaves as described before. This second point of view is an infinitesimal point of view, in contrast to the integral point of view described before. The content of the theorem of Frobenius is the equivalence of these viewpoints. For q = 1 such a system is given by a nonsingular differential form ω of degree one. The Frobenius integrability condition is then $\omega \wedge d\omega = 0$.

A singular foliation will have the decomposition property described above outside an exceptional (singular) set, frequently of zero measure. An important theme in the theory of dynamical systems is the study of singularities. The foliations discussed in these lectures rarely have singularities, though the study of singular foliations is a subject of great interest and significance.

As already stated in the preface, the geometric theory of dynamical systems was founded by Poincaré at the end of the nineteenth century. The origin of the theory of foliations lies perhaps in a question of H. Hopf (1935) on the existence of an integrable field of planes on S^3, according to a comment by Reeb 1978. ("Reeb 1978" refers to the list of books and surveys on particular aspects of foliations at the end of this Introduction). The answer is known to be affirmative: all odd-dimensional spheres admit foliations of codimension one. The Reeb foliation is an example on S^3 (see Chapter 2). The pioneers of foliation theory were Ehresmann and Reeb, the latter in particular coined the term foliation (Reeb 1952).

An idea of Poincaré in the study of dynamical systems was to analyze the intersections of an orbit with a transversal manifold. E.g. a closed orbit corresponds to a periodic point of the resulting Poincaré map (the image of a point x being the next intersection of the orbit through x with the transverse manifold). This idea finds its development in the transversal geometry of a foliation, which plays a central role in the current research on the subject. The leaves being themselves p-dimensional manifolds, there are further geometric properties pertaining to the leaves (tangential geometry of a foliation). It is fundamental to keep these complementary viewpoints of transversal and tangential geometry of a foliation in mind.

To indicate the flavor of foliation theory, here are some themes. Where do foliations naturally appear? The answer is that the world is full of them, once the mind's eye is trained to detect them. For instance the introduction of coordinates in an n-dimensional space amounts to choosing a family of n-pairwise transversal foliations of codimension one on the space. What foliations can occur on a given manifold M? What are the possible leaf types of foliations on M? Must there necessarily be compact leaves, or what are conditions having this as a consequence? Can Euclidean $\mathbb{R}^3$ be foliated by circles (a problem raised by Epstein and Millet)? How are the different leaves of a foliation related? The latter question typically pertains to transverse geometry. What is the effect of a Riemannian metric on M on the possible foliations on M? Is it possible, for instance, to find a foliation of M by geodesics (a question raised by Gluck)? More generally, can one find a foliation by minimal submanifolds? What is the effect of curvature properties of a metric on the possible foliations on M? Under which conditions can one compare a given foliation with certain standard foliations? The Bernstein problem is of this nature.

Such questions, and appropriate techniques and answers, constitute the field of (Riemannian) geometric foliation theory. In fact, ordinary manifolds always carry the trivial foliation by points (q = n), and the other extreme case is the tautological one leaf foliation (q = 0). In this sense the concept of a foliation is a natural and simultaneous generalization of the concepts of a manifold and a submersion, and is thus a natural object of study by geometric methods. For excellent surveys we refer to the two early reports by Lawson 1974, 1975 and the Ergebnisbericht by Reinhart 1983.

We mention now a few examples of foliations, that are outside the scope of these lectures, and in that sense atypical. They are nevertheless of great interest, and might indicate directions for future development of the field.

First we consider the space $\mathcal{M}$ of all Riemannian metrics on a compact and oriented manifold M. Let $\mathcal{D}$ be the group of diffeomorphisms of M. It acts on $\mathcal{M}$ by pulling back metrics, i.e. for $\varphi \in \mathcal{D}$ the metric g is sent to $\varphi^* g$. The orbits of this action define a foliation of $\mathcal{M}$. The presence of symmetries gives rise to singularities, since the isometry group of a metric gives rise to isotropy for this action. But in many respects this is still a tractable situation. For example, there is a canonical Riemannian metric on $\mathcal{M}$ for which $\mathcal{D}$ acts by isometries, a situation dicussed in these lectures in great detail in the case of absence of singularities. Some aspects of this infinite dimensional analog are discussed in a forthcoming paper by Bourguignon and the author.

Next we consider an electrostatic field, and the associated foliation by the equipotential surfaces. Here the charges produce singularities, which is of course the core of that subject.

Finally we mention the fascinating subject of liquid crystals. The pictures of liquid crystals are very suggestive of the presence of foliations. In particular the theory of defects might be susceptible to some geometric treatment. There are some preliminary contributions of Langevin to these questions.

Books and Surveys on particular aspects of foliations

Bott, R.
- 1972 Lectures on characteristic classes and foliations, Lecture Notes in Mathematics 279, 1-94, Springer Verlag, New York.
- 1973 Gelfand-Fuks cohomology and foliations, Proc. Symp. New Mexico State University.
- 1976 On characteristic classes in the framework of Gelfand-Fuks cohomology, Astérisque 32-33, 113-139.

Camacho, C. and Neto, A. L.
- 1979 Geometric theory of foliations, I.M.P.A. Rio de Janeiro [Portuguese]. Translation: Birhkäuser, Boston (1985).

Conlon, L.
- 1985 Foliations and exotic classes, Lectures at the Universidad de Extremadura, Jaranville de la Vera (Caceres).

Connes, A.
- 1982 A survey of foliations and operator algebras, Proc. Symp. Pure Math. 38, Part 1, 521-628.
- 1985 Non commutative differential geometry, Publ. Math. IHES 62, 41-144.

Ehresmann, Ch.
- 1961 Structures feuilletées, Proc. Fifth Canadian Math. Congress.

Fuks, D. B.
- 1978 Cohomology of infinite-dimensional Lie algebras and characteristic classes of foliations, Itogi Nauki-Seriya "Matematika" 10, 179-286 [Russian]. Translation: J. Soviet Math 11(1979), 922-980.
- 1981 Foliations, Itogi Nauki-Seriya Algebra, Topologiya, Geometriya 18, 151-213 [Russian]. Translation: J. Soviet Math. 18(1982), 255-291.

Haefliger, A.
- 1958 Structures feuilletées et cohomologie à valeur dans un faisceau de groupoïdes, Comment. Math. Helv. 32, 248-329.
- 1972 Sur les classes caractéristiques des feuilletages, Sém. Bourbaki 412-01 to 412-21.
- 1976 Differentiable cohomology, C.I.M.E. Lectures.

Hector, G. and Hirsch, U.
- 1981 Introduction to the geometry of foliations, Vieweg Verlag, Braunschweig, Part A.
- 1983 Part B.

Kamber, F. W. and Tondeur, Ph.
- 1975 Foliated bundles and characteristic classes, Lecture Notes in Mathematics 493, Springer Verlag, New York.
- 1978 G-foliations and their characteristic classes, Bull. Amer. Math. Soc. 84, 1086-1124.

Lawson, H. B.
- 1974 Foliations, Bull. Amer. Math. Soc. 80, 369-418.
- 1977 Lectures on the quantitative theory of foliations, CBMS Regional Conf. Series, Vol. 27.

Molino, P.
- 1983 Feuilletages riemanniens, Secrétariat des Mathématiques, Université des Sciences et Techniques du Languedoc, 1982-1983.

Reeb, G.
- 1952 Sur certaines propriétés topologiques des variétés feuilletées, Actualités Sci. Indust., Hermann, Paris.
- 1978 Structures feuilletées, Lecture Notes in Mathematics 652, 104-113, Springer Verlag, New York.

Reinhart, B. L.
- 1983 Differential geometry of foliations, Ergeb. Math. 99, Springer Verlag, New York.

Vaisman, I.
- 1973 Cohomology and differential forms, Dekker, New York.

CHAPTER 2
INTEGRABLE FORMS

In this chapter we discuss a simple, but already quite interesting special case, namely transversally oriented foliations of codimension one. The general definitions are deferred to Chapter 3. Let M^{n+1} be a smooth manifold, and $\omega \in \Omega^1(M)$ a nonsingular differential form of degree one, i.e. $\omega_x \neq 0$ for all $x \in M$. Then ω defines a field of hyperplanes $L \subset TM$ by $L_x = \ker \omega_x$. L is a smooth codimension one subbundle of TM.

An example is the differential $\omega = df$ of a smooth function $f : M \to \mathbb{R}$. The nonsingularity of ω requires the absence of critical points for f. Thus M can certainly not be compact. In this situation the level surfaces $f = \text{const}$ define a foliation of codimension one on M with tangent spaces given by the subbundle L.

A necessary condition for a 1-form ω to occur in the form $\omega = df$ is the condition $d\omega = 0$. Now if the De Rahm cohomology class $[\omega] \in H^1_{DR}(M)$ is nontrivial, there is no global solution of $\omega = df$. What is the situation if M is simply connected? Then $H^1_{DR}(M) = 0$, the equation $\omega = df$ has a global solution $f : M \to \mathbb{R}$, and the field of hyperplanes defined by $\omega = 0$ is tangent to the foliation by the level surfaces of the function f.

The multiplication of the 1-form ω by a nonzero function h does not modify the subbundle $L \subset TM$. A natural variation of the integration problem discussed above, is to ask for the local solvability of the equation $\omega = gdf$ with $f,g : U \to \mathbb{R}$, g non-zero, on an open subset U of M ($g = \frac{1}{h}$ in the notation above). Exterior differentiation yields the necessary integrability condition

$$\omega \wedge d\omega = (g\,df) \wedge d(g\,df) = g\,df \wedge dg \wedge df = 0.$$

This condition is simultaneously satisfied for ω and $h\omega$, since

$$h\omega \wedge d(h\omega) = h\omega \wedge (dh \wedge \omega + h d\omega) = h^2\, \omega \wedge d\omega.$$

The theorem of Frobenius states conversely, that the integrability condition $\omega \wedge d\omega = 0$ for a nonsingular 1-form ω guarantees the local solvability of the equation $\omega = g\,df$ with $f, g : U \to \mathbb{R}$ and g non-zero.

For $n = 2$ the 3-form $\omega \wedge d\omega$ is necessarily 0. A nonsingular 1-form defines by $\omega = 0$ a line field. If M^2 is compact, it follows that the Euler characteristic $\chi(M^2) = 0$. Thus M^2 is the torus T^2 in the orientable case, the Klein bottle K^2 in the nonorientable case.

Figure 2.1 exhibits an example of a closed 1-form on $T^2 = \mathbb{R}^2/\mathbb{Z}^2$ and the corresponding line field.

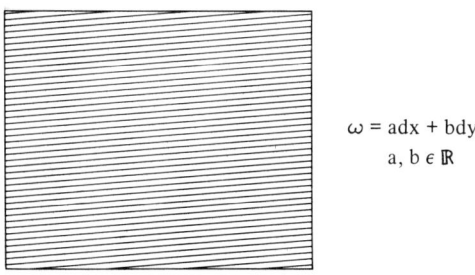

Figure 2.1

Another possibility for a line field is indicated in Figure 2.2 by its integral curves

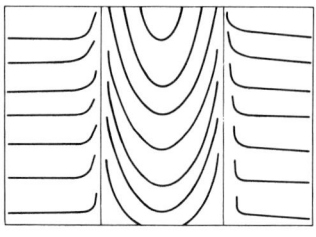

Figure 2.2

In cylindrical coordinates on $\mathbb{R}^3$ consider the 1-form

$$\omega = h(r)dr + (1 - h(r))dz$$

where $h(r)$ is a smooth monotone non-decreasing function with $h(r) = 0$ for $r \leq 0$, $0 < h(r) < 1$ for $0 < r < 1$, and $h(r) = 1$ for $r \geq 1$. Then $d\omega = -h'dr \wedge dz$ and thus $\omega \wedge d\omega = 0$. The resulting foliation of the cylinder $D^2 \times \mathbb{R}$ gives a foliation of the solid torus $D^2 \times S^1$. The <u>Reeb foliation</u> of S^3 is obtained from two copies of $D^2 \times S^1$ by glueing them together along their common boundary $S^1 \times S^1$. The 2-torus appears thus as the unique closed ($\equiv$ compact without boundary) leaf of this foliation.

A theorem of Novikov states that every foliation of codimension one on a closed 3-manifold with finite fundamental group has a closed leaf. In fact Novikov proves that every such foliation has a Reeb component, i.e. a subset which is a union of leaves and which is diffeomorphic to $D^2 \times S^1$ with the foliation described above. The closed boundary leaf is thus a torus [N]. (This refers to the entry [N] in the list of references following Chapter 13).

Note that the Reeb foliation on S^3 cannot be defined by a closed 1-form ω. Since $H^1_{DR}(S^3) = 0$, such a form would be df for some $f : S^3 \to \mathbb{R}$. At the critical points of f, ω would then have to vanish.

This argument visibly applies to any codimension one foliation on a closed manifold with finite fundamental group.

A further example is given by a force field in $\mathbb{R}^3$ defined by a 1-form ω. If the force field derives from a potential $\Phi : \mathbb{R}^3 \to \mathbb{R}$, then $\omega = -d\Phi$ (with the sign usual in physics). The equipotential surfaces define a foliation of codimension one. The force or field lines are orthogonal trajectories.

Consider the phase space M (in $\mathbb{R}^k$) of a thermodynamical system. The pressure P, volume V and internal energy E are functions on M. The external work provided by a change of state along a path $\gamma : [0,1] \to M$ is obtained by the line integral $\int_\gamma \omega$, where $\omega = dE + PdV \in \Omega^1(M)$. This is the content of the first principle of thermodynamics (energy conservation). As observed by Caratheodory [C], the second principle of thermodynamics is precisely the integrability condition $\omega \wedge d\omega = 0$. By Frobenius, it is equivalent to the local solvability of $\omega = T\,dS$ (this defines the temperature T and the entropy S). This yields all thermodynamical identities. E.g. differentiation of $\frac{1}{T}\omega = dS$ for $\omega = dE + PdV$ yields

$$0 = d\left[\frac{1}{T}\omega\right] = -\frac{1}{T^2}dT \wedge \omega + \frac{1}{T}dP \wedge dV,$$

and with $E = E(T,V)$, $P = P(T,V)$ the identity

$$0 = -\frac{1}{T^2}dT \wedge \left[\frac{\partial E}{\partial V} + P\right]dV + \frac{1}{T}\frac{\partial P}{\partial T}dT \wedge dV,$$

which in turn implies

$$\frac{\partial E}{\partial V} = T \frac{\partial P}{\partial T} - P,$$

an identity well known in thermodynamics.

For a codimension one foliation on a closed manifold M, the existence of a transversal line field shows that the Euler characteristic $\chi(M)$ vanishes. A theorem of Thurston states that conversely every closed manifold M with $\chi(M) = 0$ admits a codimension one foliation [TH 2].

It is important to note that these results all concern smooth foliations. Haefliger has shown that on a closed simply connected manifold no real analytic codimension one foliation exists [HA 2].

We return to a discussion of the integrability condition.

2.1. PROPOSITION. <u>Let</u> M <u>be a smooth manifold, and</u> $\omega \in \Omega^1(M)$ <u>a nonsingular</u> 1-<u>form</u>. <u>The following conditions are equivalent</u>:
(i) $\omega \wedge d\omega = 0$;
(ii) $\omega[X,Y] = 0$ <u>for all vector fields</u> X,Y <u>with</u> $i(X)\omega = 0$, $i(Y)\omega = 0$;
(iii) $d\omega = \alpha \wedge \omega$ <u>for some</u> $\alpha \in \Omega^1(M)$.

PROOF. (i) => (iii): First we prove the local existence of α. Let U be an open set trivializing the dual tangent bundle $T^*M|U$. The nonsingular 1-form $\omega = \omega_1$ can then be completed to a basis $\omega_1,\ldots,\omega_{n+1}$ of $\Omega^1(U)$, where n+1 = dim M. Thus $d\omega = \sum_{j<k} b_{jk} \omega_j \wedge \omega_k$ and by assumption

$$\omega \wedge d\omega = \omega_1 \wedge (\sum_{j<k} b_{jk} \omega_j \wedge \omega_k) = 0.$$

The 3-forms $\omega_1 \wedge \omega_j \wedge \omega_k$ for $1 < j < k$ are linearly independent, so that $b_{jk} = 0$ for $1 < j < k$. It follows that on U

$$d\omega = \sum_{k=2}^{n+1} b_{1k}\, \omega_1 \wedge \omega_k = (-\sum_{k=2}^{n+1} b_{1k}\, \omega_k) \wedge \omega_1.$$

Thus with $\alpha = -\sum_{k=2}^{n+1} b_{1k}\, \omega_k$, we have $d\omega = \alpha \wedge \omega$ on U.

The global existence of α follows from a partition of unity argument. Let $\mathcal{U} = \{U_\gamma\}$ be an open covering of M by subsets U_γ trivializing $T^*M|U_\gamma$ and $\lambda = \{\lambda_\gamma\}$ a smooth partition of 1 subordinate to $\mathcal{U}$. On each U_γ we obtain $\alpha_\gamma \in \Omega^1(U_\gamma)$, such that $d\omega = \alpha_\gamma \wedge \omega$ holds on U_γ. Define $\alpha = \sum_\gamma \lambda_\gamma \in \Omega^1(M)$. Then

$$\alpha \wedge \omega = \sum_\gamma \lambda_\gamma (\alpha_\gamma \wedge \omega) = \sum_\gamma \lambda_\gamma \, d\omega|U_\gamma = \sum_\gamma \lambda_\gamma \, d\omega = d\omega$$

which proves (iii).

(iii) => (ii): For $i(X)\omega = 0$, $i(Y)\omega = 0$ we have

$$d\omega(X,Y) = -\omega[X,Y] \quad \text{and} \quad (\alpha \wedge \omega)(X,Y) = 0.$$

Thus $d\omega = \alpha \wedge \omega$ implies (ii). Only the existence of a local form α is needed in this argument.

(ii) => (i): Again locally we extend ω to a basis $\omega_1,\ldots,\omega_{n+1}$ of $\Omega^1(V)$, with $\omega = \omega_1$. Let $X_1,\ldots,X_{n+1}$ be the dual frame of $TM|U$. Then $d\omega = \sum b_{jk}\, \omega_j \wedge \omega_k$ and

$$b_{jk} = d\omega(X_j, X_k) = -\omega[X_j, X_k].$$

Thus by assumption $b_{jk} = 0$ for $1 < j < k$. It follows that

$$d\omega = \sum_{k=2}^{n+1} b_{1k}\, \omega \wedge \omega_k, \quad \text{and therefore} \quad \omega \wedge d\omega = 0. \quad \blacksquare$$

2.2 REMARK. In general, the 1-form α in (iii) is not even locally well defined. It is however related to the (local) solvability of $\omega = gdf$ in the following way. This identity implies $d\omega = dg \wedge df$, which can be written

$$d\omega = \frac{1}{g}\, dg \wedge gdf = d\,\log|g| \wedge \omega,$$

so that $\alpha = d\,\log|g|$ is of the desired form.

To what extent is α globally determined? Let α', α both be 1-forms with $d\omega = \alpha' \wedge \omega = \alpha \wedge \omega$. Then $(\alpha' - \alpha) \wedge \omega = 0$, i.e. the 1-forms $\alpha' - \alpha$ and ω are linearly dependent. It follows that $\alpha' - \alpha = f\omega$ for some function $f : M \to \mathbb{R}$. This describes the indeterminacy in α.

Next we give a geometric construction of α.

2.3 PROPOSITION. <u>Let</u> Z <u>be a vector field on</u> M <u>satisfying</u> $i(Z)\omega = 1$, <u>e.g. a vector field orthogonal to the hyperplanes defined by</u> $\omega = 0$, <u>with respect to a Riemannian metric</u> g_M <u>on</u> M, <u>and scaled so as to have</u> $i(Z)\omega = 1$. <u>Then the form</u> $\alpha = -\theta(Z)\omega$ <u>is a proper choice, i.e.</u> $d\omega = -\theta(Z)\omega \wedge \omega$.

PROOF. The 2-forms $-\theta(Z)\omega \wedge \omega$ and $d\omega$ both vanish for X,Y with $i(X)\omega = 0$, $i(Y)\omega = 0$. Namely

$$(-\theta(Z)\omega \wedge \omega)(X,Y) = (-\theta(Z)\omega)(X)\cdot\omega(Y) - (-\theta(Z)\omega)(Y)\cdot\omega(X) = 0$$

and

$$d\omega(X,Y) = X\cdot\omega(Y) - Y\cdot\omega(X) - \omega[X,Y] = 0.$$

To prove their equality, it suffices therefore to prove for any X

$$i(X)i(Z)(-\theta(Z)\omega \wedge \omega) = i(X)i(Z)d\omega \; .$$

Now

$$(i(Z)(-\theta(Z)\omega \wedge \omega) = -i(Z)\theta(Z)\omega\cdot\omega + \theta(Z)\omega\cdot i(Z)\omega.$$

But $[\theta(Z),i(Z)] = i[Z,Z] = 0$, so that $i(Z)\theta(Z)\omega = \theta(Z)i(Z)\omega = Z1 = 0$ (the formulas of the exterior calculus, together with degree and sign conventions, are summarized at the end of this chapter). It follows that

$$i(Z)(-\theta(Z)\omega \wedge \omega) = \theta(Z)\omega = i(Z)d\omega.$$

This implies the desired equality. ∎

The existence of the global form α leads to the following construction of Godbillon and Vey.

2.4 THEOREM [GV]. Let M be a smooth manifold, $\omega \in \Omega^1(M)$ a nonsingular integrable 1-form, and α a global 1-form as in property (iii) of Proposition 2.1. Then:

(i) $d(\alpha \wedge d\alpha) = 0$,

(ii) the De Rham cohomology class $[\alpha \wedge d\alpha] \in H^3_{DR}(M)$ is independent of the choice of α satisfying $d\omega = \alpha \wedge \omega$ (this class is called the Godbillon-Vey class of ω);

(iii) the De Rham cohomology class associated to ω and $h\omega$ with non-zero h is the same.

In the proof we use the following local result (already exploited before for the case $\beta = d\omega$).

2.5 LEMMA. Let β be a 2-form and ω a nonsingular 1-form. If $\beta \wedge \omega = 0$, then (locally) $\beta = \gamma \wedge \omega$ for some 1-form γ.

PROOF. Complete (locally) $\omega = \omega_1$ to a basis $\omega_1, \ldots, \omega_{n+1}$ of 1-forms. Then $\beta = \sum_{j<k} b_{jk} \omega_j \wedge \omega_k$. The condition $\beta \wedge \omega = 0$ implies again that $b_{jk} = 0$ for $1 < j < k$ and therefore

$$\beta = \sum_{k=2}^{n+1} b_{1k} \omega_1 \wedge \omega_k = \gamma \wedge \omega$$

with $\gamma = - \sum_{k=2}^{n+1} b_{1k} \omega_k$. ∎

PROOF OF THEOREM 2.4. Differentiation of $d\omega = \alpha \wedge \omega$ yields

$$0 = d\alpha \wedge \omega - \alpha \wedge d\omega = d\alpha \wedge \omega,$$

and similarly $d\alpha' \wedge \omega = 0$.

(i) Applying Lemma 2.5, there is locally a 1-form γ with $d\alpha = \gamma \wedge \omega$. Therefore

$$d(\alpha \wedge d\alpha) = d\alpha \wedge d\alpha = (\gamma \wedge \omega) \wedge (\gamma \wedge \omega) = 0.$$

(ii) Let α, α' be 1-forms such that $d\omega = \alpha \wedge \omega = \alpha' \wedge \omega$. Then $\alpha' = \alpha + f\omega$, so $\alpha' \wedge d\alpha' = \alpha \wedge d\alpha' + f\omega \wedge d\alpha'$. It follows that

$$\alpha' \wedge d\alpha' = \alpha \wedge d\alpha + \alpha \wedge d(f\omega).$$

Further

$$d(\alpha \wedge f\omega) = d\alpha \wedge f\omega - \alpha \wedge d(f\omega) = -\alpha \wedge d(f\omega).$$

Thus

$$\alpha' \wedge d\alpha' = \alpha \wedge d\alpha - d(\alpha \wedge f\omega),$$

which proves (ii).

(iii) Let $\omega' = h\omega$ with nonzero h. Then

$$d\omega' = dh \wedge \omega + h d\omega = \frac{1}{h} dh \wedge h\omega + h\alpha \wedge \omega = (d \log |h| + \alpha) \wedge \omega',$$

so that $\alpha' = \alpha + d \log |h|$ can be used for the definition of the Godbillon-Vey class of ω'. But then

$$\alpha' \wedge d\alpha' = (\alpha + d \log |h|) \wedge d\alpha = \alpha \wedge d\alpha + d(\log |h| \wedge d\alpha),$$

which proves (iii) ∎

Consider the canonical 1-form $d\theta$ on S^1. A submersion $f : M \to S^1$ defines a closed 1-form $\omega = f^*d\theta$ on M. The corresponding foliation of M by the fibers of f is therefore an example of a foliation with trivial Godbillon-Vey class (in fact by construction even trivial Godbillon-Vey form, since $\alpha = 0$ is a proper choice).

The first example of a nontrivial Godbillon-Vey class was given by Roussarie. The idea is to consider the foliation of a Lie group G by the left cosets gH of a connected subgroup $H \subset G$. Let Γ be a discrete subgroup of G with orbit space $\Gamma \backslash G$ (left action). The foliation of G by H defines a foliation of $\Gamma \backslash G$ which is a compact space for co-compact Γ. An interesting case is $G = PSL(2,\mathbb{R})$, the quotient of $SL(2,\mathbb{R})$ by the subgroup $\pm$ id. This is the group of complex automorphisms of the unit disc D. Let M_g be a closed Riemannian surface of genus $g > 1$ and fundamental group Γ_g. The universal covering $\tilde{M}_g \cong D$, and Γ_g appears via deck transformations as a subgroup of G. But Aut (D) acts simply transitively on the unit tangent bundle T_1D, so that $G \cong$ Aut $(D) \cong T_1D \cong T_1\tilde{M}_g$, and $\Gamma_g \backslash G \cong \Gamma_g \backslash T_1\tilde{M}_g \cong T_1M_g$, the unit tangent bundle of M_g. A basis of the Lie algebra $\mathfrak{g}$ of G consists of the matrices $A = \begin{bmatrix} 1/2 & 0 \\ 0 & -1/2 \end{bmatrix}$, $B^+ = \begin{bmatrix} 0 & 1 \\ 0 & 0 \end{bmatrix}$, $B^- = \begin{bmatrix} 0 & 0 \\ 1 & 0 \end{bmatrix}$. The subalgebra $\mathfrak{h}$ with basis A, B^+ defines a connected subgroup $H \subset G$. The corresponding codimension one foliation of $\Gamma \backslash G = T_1M_g$ has a nonzero Godbillon-Vey number [GV].

To verify this fact, consider the dual basis μ, ν^+, ν^- of $\mathfrak{g}^*$. Using the Lie algebra relations $[A,B^+] = B^+$, $[A,B^-] = -B^-$, $[B^+,B^-] = 2A$ one verifies the identity

$$d\nu^- = \mu \wedge \nu^- .$$

Thus the integrable 1-form ν^- defines the foliation of G by H. Its Godbillon-Vey form is $\mu \wedge d\mu$. One finds that

$$\mu \wedge d\mu = -2\mu \wedge \nu^+ \wedge \nu^- ,$$

which is a left invariant volume form on G. Thus it defines a volume form on the quotient $\Gamma \setminus G$. It follows that its integral over the compact manifold $\Gamma_g \setminus G$ is a nonzero number. Up to the factor $4\pi^2$, this is in fact the Euler characteristic

$$\chi(M_g) = 2 - 2g \quad (g > 1) .$$

As in the example just discussed, the Godbillon-Vey class is of special interest for integrable 1-forms ω on closed oriented 3-manifolds. Its integral over the fundamental cycle defines then the Godbillon-Vey number of ω. For S^3 Thurston showed that every real number is realized as a Godbillon-Vey number in this fashion [TH 1]. The Godbillon-Vey number of the Reeb foliation can be shown to be zero. Unpublished work of Duminy establishes that the Godbillon-Vey class vanishes if all the leaves of a codimension one foliation are manifolds with nonexponential growth. Recent papers discussing the significance of the Godbillon-Vey invariant are [CC] and [HK 2].

The formula for α in 2.3 yields the following result (Thurston).

2.6 PROPOSITION. Let Z be a vector field with $i(Z)\omega = 1$. Then the Godbillon-Vey class of the foliation defined by ω is represented by the form

$$- \omega \wedge \theta(Z)\omega \wedge \theta(Z)^2\omega.$$

PROOF. $d\alpha = d(-\theta(Z)\omega) = -\theta(Z)d\omega = -\theta(Z)(-\theta(Z)\omega \wedge \omega = \theta(Z)^2\omega \wedge \omega.$

This yields the desired formula for $\alpha \wedge d\alpha$. ∎

As an illustration consider the flow φ_t generated by Z. If $\varphi_t^*\omega = \omega$ for all t, then

$$\theta(Z)\omega = \frac{d}{dt}\Big|_{t=0}\varphi_t^*\omega = 0,$$

so the Godbillon-Vey form of ω vanishes. In fact under this hypothesis we have even necessarily $d\omega = 0$ by Proposition 2.3. Conversely $d\omega = 0$ implies $\theta(Z)\omega = di(Z)\omega = 0$. So this happens precisely for closed ω.

Next we give a geometric interpretation of the Godbillon-Vey construction in terms of a connection and its curvature form. For this purpose let Q^* denote the bundle of multiples of the form ω (the reason for the $*$ will appear in Chapter 3, where we define a bundle Q, of which Q^* is the dual bundle). Since ω is nonsingular, $Q^* \subset T^*M$ is a line bundle in the cotangent bundle. The form ω is a trivializing section of Q^*. A connection in Q^* is therefore completely determined by the formulas

$$\nabla_X \omega = \alpha(X) \cdot \omega$$

$$\nabla_X(f\omega) = Xf \cdot \omega + f \cdot \nabla_X \omega$$

for vector fields X and functions f. It is immediate to verify that ∇ satisfies all properties required of a linear connection. Let R be the curvature tensor of ∇.

2.7 PROPOSITION. <u>For vector fields</u> X,Y <u>we have a bundle map</u> $R(X,Y) : Q^* \longrightarrow Q^*$, <u>which is given by</u> $R(X,Y)\omega = d\alpha(X,Y) \cdot \omega$.

PROOF. $R(X,Y)\omega = \nabla_X \nabla_Y \omega - \nabla_Y \nabla_X \omega - \nabla_{[X,Y]} \omega = X\alpha(Y) \cdot \omega + \alpha(Y) \cdot \alpha(X) \cdot \omega - Y\alpha(X) \cdot \omega - \alpha(X) \cdot \alpha(Y) \cdot \omega - \alpha[X,Y] \cdot \omega = d\alpha(X,Y) \cdot \omega$ ∎

2.8 REMARK. Assuming $i(X)\omega = 0$ and $i(Y)\omega = 0$, it follows that $d\alpha(X,Y) = 0$. Namely $d\omega = \alpha \wedge \omega$ implies $0 = d\alpha \wedge \omega$, hence locally $d\alpha = \gamma \wedge \omega$. Thus

$$i(X)\omega = 0 \Rightarrow i(X)d\alpha = i(X)\gamma \cdot \omega$$

and thus $i(Y)i(X)d\alpha = 0$, if $i(Y)\omega = 0$. This implies by 2.7 that

$$i(X)\omega = 0, \ i(Y)\omega = 0 \Rightarrow R(X,Y) = 0.$$

In other words the curvature is zero along the leaves of the foliation defined by ω.

In view of the observations above, the Godbillon-Vey form $\alpha \wedge d\alpha$ is therefore the exterior product of the connection form α and the corresponding curvature 2-form. This is in contrast to the Chern-Weil construction, where differential forms are constructed from the curvature form alone. This discussion hints at the generalized Godbillon-Vey construction, using more general products of connection and curvature forms. See [BO 2] [HA 4] [KT 3] for discussions of this topic.

A final remark on the topic of integrable 1-forms. The Frobenius condition $\omega \wedge d\omega = 0$ is equivalent to the local solvability of $\omega = gdf$ with nonzero g. Assume this equation to hold globally. Then with $\alpha = d \log |g|$ we have

$$d\omega = dg \wedge df = \frac{1}{g} dg \wedge gdf = \alpha \wedge \omega$$

and clearly $d\alpha = 0$. In particular the Godbillon-Vey class is necessarily zero. In other words for an integrable form ω with nontrivial Godbillon-Vey class, no global integrating factor g can exist.

We conclude this chapter with a short review of FORMULAS AND CONVENTIONS FOR THE EXTERIOR CALCULUS.

The differential forms $\Omega^{\cdot}(M)$ are a (graded) commutative algebra:

$$\alpha \wedge \beta = (-1)^{\deg \alpha \cdot \deg \beta} \beta \wedge \alpha.$$

A derivation $D : \Omega^{\cdot}(M) \to \Omega^{\cdot + r}$ of degree r satisfies the rule

$$D(\alpha \wedge \beta) = D\alpha \wedge \beta + (-1)^{\deg \alpha \cdot r} \alpha \wedge D\beta.$$

The fundamental operations are the exterior derivative, interior product and Lie derivative, which are, respectively

$$d : \Omega^{\cdot}(M) \to \Omega^{\cdot+1}(M), \quad \text{a derivation of degree } 1;$$

$$i(X) : \Omega^{\cdot}(M) \to \Omega^{\cdot-1}(M), \quad \text{a derivation of degree } -1;$$

$$\theta(X) : \Omega^{\cdot}(M) \to \Omega^{\cdot}(M), \quad \text{a derivation of degree } 0.$$

The following identities hold:

$$\begin{aligned} d^2 &= 0 \\ i(X)^2 &= 0 \\ \theta(X) &= di(X) + i(X)d \quad \text{(Cartan formula)} \\ \theta[X,Y] &= [\theta(X),\theta(Y)] \\ [\theta(X),i(Y)] &= i[X,Y] \,. \end{aligned}$$

In the last two formulas, the following convention is used. For derivations D, D' of degree r, r' respectively, the commutator is the derivation of degree $r + r'$ defined by

$$[D,D'] = DD' + (-1)^{rr'+1} D'D.$$

CHAPTER 3

FOLIATIONS

Let M be a smooth manifold and $n = \dim M$. <u>A foliation</u> $\mathcal{F}$ <u>on</u> M <u>of</u> <u>dimension</u> p <u>and codimension</u> q $(p + q = n)$ is a partition $\{\mathcal{L}_\alpha\}_{\alpha \in A}$ of M into connected subsets with the following property. For every point of M there is an open neighborhood U and a chart (x,y)

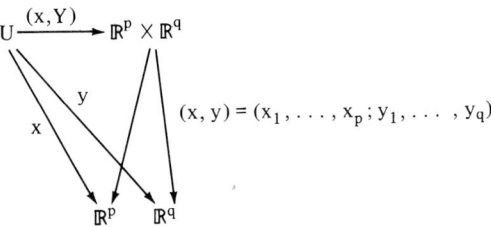

Figure 3.1

such that for each leaf $\mathcal{L}_\alpha$ the connected components of $U \cap \mathcal{L}_\alpha$ are defined by the equations $y = \text{const}$, i.e. $y_1 = \text{const}, \ldots, y_q = \text{const}$.

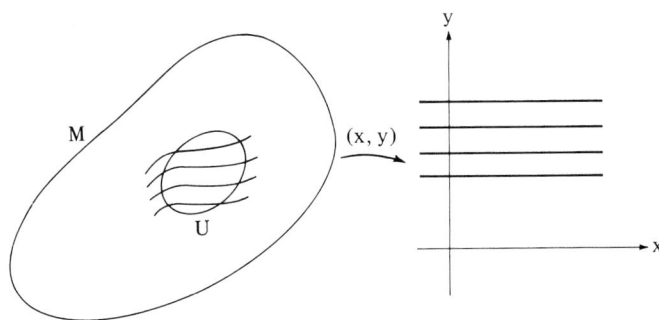

Figure 3.2

Such a chart is a distinguished chart.

The connected components of the sets y = constant in a distinguished chart are called plaques (plates) of $\mathcal{F}$. Fixing y, the map x → (x,y) is a smooth embedding. Thus the plaques are connected p-dimensional submanifolds of M. This shows that each leaf $\mathcal{L}_\alpha$ is a p-dimensional connected submanifold of M. It is a union of plaques. Note that the natural manifold topology on $\mathcal{L}_\alpha$ is not necessarily the topology induced from M.

An example of a foliation is given by a smooth submersion f : M → Y^q to a q-dimensional manifold. The connected components of the inverse images of the points of Y define a codimension q foliation of M. Note that by its very definition, locally a foliation is defined by a submersion. But this need not be the case globally. As an example take the Kronecker line on the 2-torus $T^2 = \mathbb{R}^2/\mathbb{Z}^2$ defined by the form ω = adx + bdy with a,b constant and $\frac{b}{a} \notin \mathbb{Q}$. Every leaf is dense in T^2, so the foliation cannot be given by a submersion.

The distinguished charts represent a microscopic view of a foliation. A macroscopic view is encoded in the neighborhood relations between overlapping charts.

Let $\varphi_\alpha = (x_\alpha, y_\alpha) : U_\alpha \to \mathbb{R}^n$, $\varphi_\beta = (x_\beta, y_\beta) : U_\beta \to \mathbb{R}^n$ be distinguished charts for a foliation $\mathcal{F}$ on M. Then the leaves in $U_\alpha \cap U_\beta$ are given by y_α = const and y_β = const. This shows that in the neighborhood relations $x_\beta = x_\beta(x_\alpha, y_\alpha)$, $y_\beta = y_\beta(x_\alpha, y_\alpha)$ in fact the y_β do not depend on x_α. Thus $y_\beta = \gamma_{\beta\alpha}(y_\alpha)$, where $\gamma_{\beta\alpha}$ is a local diffeomorphism of $\mathbb{R}^q$, and is called a transition function. Let f_α be the composition $U_\alpha \to \mathbb{R}^n \xrightarrow{proj} \mathbb{R}^q$ and similarly for f_β. Then f_α, f_β are related by

(3.1) $$f_\beta = \gamma_{\beta\alpha} f_\alpha \quad \text{in} \quad U_\alpha \cap U_\beta.$$

This leads to the equivalent definition of a foliation by a family of submersions $f_\alpha : U_\alpha \to \mathbb{R}^q$ for an open covering $\mathcal{U} = \{U_\alpha, U_\beta, \ldots\}$, together with a family of local diffeomorphisms (transition functions) $\gamma_{\beta\alpha}$ of $\mathbb{R}^q$, for each (α,β) with $U_\alpha \cap U_\beta \neq \emptyset$, satisfying the relations (3.1). For these relations to be compatible on triple intersections, one needs the cocycle conditions

(3.2) $$\gamma_{\delta\alpha} = \gamma_{\delta\beta} \circ \gamma_{\beta\alpha} \text{ on } U_\alpha \cap U_\beta \cap U_\delta \neq \emptyset.$$

The tangent bundle L of a foliation is the subbundle of TM, consisting of all vectors tangent to the leaves of $\mathcal{F}$. This is a p-dimensional vector bundle. It is integrable or involutive in the sense that

(3.3) $$X, Y \in \Gamma(U,L) \Rightarrow [X,Y] \in \Gamma(U,L)$$

where $\Gamma(U,L)$ denotes the sections of L over an open subset U. It is convenient to simply write $X \in \Gamma L$, leaving the specification of the domain to the context.

Conversely let $L \subset TM$ be a subbundle of codimension q. It is the content of the theorem of Frobenius that if L is integrable then it is the tangent bundle of a codimension q foliation $\mathcal{F}$ on M. The leaves of $\mathcal{F}$ are constructed as the maximal connected integral submanifolds of the subbundle $L \subset TM$. The normal bundle Q of a codimension q foliation $\mathcal{F}$ on M is the quotient bundle $Q = TM/L$. Equivalently Q appears in the exact sequence of vector bundles

(3.4) $$0 \to L \to TM \xrightarrow{\pi} Q \to 0.$$

The exactness of this sequence is equivalent to the exactness of the sequence of dual vector bundles

(3.5) $$0 \to Q^* \to T^*M \to L^* \to 0.$$

In terms of a local framing $\omega_1, \ldots, \omega_q$ of Q^* on a trivializing set U the integrability condition (3.3) for L translates to the following integrability condition for Q^*:

(3.6) $$d\omega_i = \sum_{j=1}^{q} \alpha_{ij} \wedge \omega_j \quad \text{for some} \quad \alpha_{ij} \in \Omega^1(U), \; i,j = 1,\ldots,q.$$

Equivalent forms of this condition are as follows.

3.7 PROPOSITION. Let M be a smooth manifold, U an open subset of M, and $\omega_1, \ldots, \omega_q \subset \Omega^1(U)$ linearly independent forms at every point $x \in U$. Then the following conditions are equivalent:

(i) $d\omega_i = \sum_{j=1}^{q} \alpha_{ij} \wedge \omega_j$ locally for some 1-forms α_{ij}, $i,j = 1,\ldots,q$;

(ii) there exists a 1-form $\alpha \in \Omega^1(U)$ such that
$$d(\omega_1 \wedge \ldots \wedge \omega_q) = \alpha \wedge \omega_1 \wedge \ldots \wedge \omega_q;$$

(iii) $\omega_1 \wedge \ldots \wedge \omega_q \wedge d\omega_i = 0$ $(i = 1,\ldots,q)$.

PROOF. (i) => (ii): With $\alpha = \sum_{j=1}^{q} \alpha_{jj}$ we have

$$d(\omega_1 \wedge \ldots \wedge \omega_q) = \sum_{j=1}^{q} (-1)^{j-1} \omega_1 \wedge \ldots \wedge d\omega_j \wedge \ldots \wedge \omega_q = \sum_{j=1}^{q} \alpha_{jj} \omega_1 \wedge \ldots \wedge \omega_q = \alpha \wedge \omega_1 \wedge \ldots \wedge \omega_q.$$

This proves the existence of α for every open subset of U on which (i) holds. A partition of unity argument gives such a 1-form α on all of U.

(ii) => (iii): Clearly $\alpha \wedge \omega_1 \wedge \ldots \wedge \omega_q \wedge \omega_i = 0$. But by (ii)

$$\alpha \wedge \omega_1 \wedge \ldots \wedge \omega_q \wedge \omega_i = d(\omega_1 \wedge \ldots \wedge \omega_q) \wedge \omega_i$$

$$= \sum_j (-1)^{j-1} \omega_1 \wedge \ldots \wedge d\omega_j \wedge \ldots \wedge \omega_q \wedge \omega_i = (-1)^{i-1+q-i} d\omega_i \wedge \omega_1 \wedge \ldots \wedge \omega_q$$

$$= (-1)^{q-1} d\omega_i \wedge \omega_1 \wedge \ldots \wedge \omega_q.$$

Thus the latter form vanishes.

(iii) => (i): On a trivializing set U for T^*M complete $\omega_1, \ldots, \omega_q$ by $\omega_{q+1}, \ldots, \omega_n$ to a basis of $\Omega^1(U)$. Then

$$d\omega_i = \sum_{j<k} b_{jk}^i \omega_j \wedge \omega_k = 0.$$

(iii) implies then

$$\omega_1 \wedge \ldots \wedge \omega_q \wedge \sum_{j<k} b_{jk}^i \omega_j \wedge \omega_k = 0.$$

But $\omega_1 \wedge \ldots \wedge \omega_q \wedge \omega_j \wedge \omega_k$ for $q < j < k$ are linearly independent, and thus $b_{jk}^i = 0$ for $q < j < k$. Hence

$$d\omega_i = \sum_{j=1}^{q} \left[\sum_{k=j+1} - b_{jk}^{i} \omega_k \right] \wedge \omega_j = \sum_{j=1}^{q} \alpha_{ij} \wedge \omega_j$$

which proves (i). ∎

3.8 DEFINITION. <u>The foliation</u> $\mathcal{F}$ <u>is transversally orientable, if</u> $\Lambda^q Q^*$ <u>is trivial, i.e. admits a nowhere zero section</u>.

Note that the exterior power $\Lambda^q Q^*$ is a line bundle.

The codimension one foliations considered in Chapter 2 are transversally orientable, since the defining integrable 1-form is a nowhere zero section of Q^*. The situation for arbitrary q is clarified as follows.

3.9 PROPOSITION. <u>Let</u> $\mathcal{F}$ <u>be a transversally orientable foliation of codimension</u> q. <u>Then there is a nonsingular global</u> q-<u>form</u> $\nu \in \Gamma \Lambda^q Q^*$ <u>such that locally</u> ν <u>is of the form</u> $\nu = \omega_1 \wedge \ldots \wedge \omega_q$, <u>where</u> $\omega_1, \ldots, \omega_q \in \Gamma Q^*$, <u>and the tangent bundle</u> $L \in TM$ <u>is defined by</u> $\nu = 0$.

This is obvious in a distinguished chart. The proof of the global statement follows from a partition of unity argument. Note that the transition function of the line bundle $\Lambda^q Q^*$ is $\det g_{\alpha\beta}$, if $g_{\alpha\beta}$ is the transition function of Q^*. The transversal orientability assumption on $\mathcal{F}$ means that $\det g_{\alpha\beta}$ can be assumed positive throughout.

Under these assumptions one can find as in the proof of Proposition 2.1 a global 1-form α (not uniquely defined) such that

(3.10) $$d\nu = \alpha \wedge \nu.$$

Theorem 2.4 generalizes then as follows.

3.11 THEOREM (Godbillon-Vey). <u>Let</u> M <u>be a smooth manifold, and</u> $\mathcal{F}$ <u>a transversally orientable foliation of codimension</u> q. <u>Then:</u>

(i) $d(\alpha \wedge (d\alpha)^q) = 0 \in \Omega^{2q+1}(M)$;

(ii) <u>the De Rham cohomology class</u> $[\alpha \wedge (d\alpha)^q] \in H^{2q+1}_{DR}(M)$ <u>is independent of the choice of the transversal orientation</u> ν, <u>and independent of the choice of the</u> 1-<u>form</u> α <u>that satisfies</u> (3.10).

PROOF. During the proof we repeatedly use the following facts. If a 1-form β satisfies $\beta \wedge \nu = 0$, then locally

$$(3.12) \quad \beta = \sum_{j=1}^{q} b_j \omega_j, \quad \text{where} \quad \omega_1, \ldots, \omega_q \quad \text{locally frame} \quad Q^*.$$

If γ is a 1-form such that $d\gamma \wedge \nu = 0$, then locally

$$(3.13) \quad d\gamma = \sum_{j=1}^{q} \gamma_j \wedge \omega_j.$$

These facts are proved as in Lemma 2.5.

To prove (i), observe that by (3.10)

$$0 = d\alpha \wedge \nu + (-1)^q \alpha \wedge d\nu = d\alpha \wedge \nu.$$

Applying (3.13) we find

$$d\alpha = \sum_{j=1}^{q} \alpha_j \wedge \omega_j,$$

and thus

$$d(\alpha \wedge (d\alpha)^q) = (d\alpha)^{q+1} = \Big(\sum_{j=1}^{q} \alpha_j \wedge \omega_j\Big)^{q+1}$$

contains in each term $(q+1)$-fold exterior products of forms ω_j, which all vanish since $\Lambda^{q+1}\mathbf{Q}^* = 0$.

To prove (ii), consider first two transversal orientations ν, ν'. Then $\nu' = h\nu$ with a nowhere zero function h. It follows that

$$d\nu' = dh \wedge \nu + h \cdot \alpha \wedge \nu = (d \log |h| + \alpha) \wedge \nu',$$

and $\alpha' = \alpha + d \log |h|$ satisfies (3.10) for ν'. Thus

$$\alpha' \wedge (d\alpha')^q = (\alpha + d \log |h|) \wedge (d\alpha)^q = \alpha \wedge (d\alpha)^q + d(\log |h| \wedge (d\alpha)^q)$$

and $[\alpha' \wedge (d\alpha')^q] = [\alpha \wedge (d\alpha)^q]$.

Finally let α', α both satisfy (3.10). Then $\beta = \alpha' - \alpha$ satisfies $\beta \wedge \nu = 0$ and thus (3.12) holds. Further α' satisfies $d\alpha' \wedge \nu = 0$ (differentiate (3.10)), so (3.13) holds for $d\alpha'$. It follows that

$$(3.14) \qquad \beta \wedge (d\alpha')^q = \Big(\sum_{j=1}^{q} b_j \omega_j\Big) \wedge \Big(\sum_{k=1}^{q} \gamma_k \wedge \omega_k\Big)^q = 0,$$

since every term contains $(q+1)$-fold exterior products of ω_j's. This shows that

$$\alpha' \wedge (d\alpha')^q = (\alpha + \beta) \wedge (d\alpha')^q = \alpha \wedge (d\alpha')^q.$$

We calculate

$$(d\alpha')^q = (d\alpha + d\beta)^q = (d\alpha)^q + \sum_{k=1}^{q} \begin{bmatrix} q \\ k \end{bmatrix} (d\alpha)^{q-k} \wedge (d\beta)^k.$$

Thus $\alpha' \wedge (d\alpha')^q$ and $\alpha \wedge (d\alpha)^q$ differ by

(3.15)
$$\sum_{k=1}^{q} \begin{bmatrix} q \\ k \end{bmatrix} \alpha \wedge (d\alpha)^{q-k} \wedge (d\beta)^k.$$

But

$$\alpha \wedge (d\alpha)^{q-k} \wedge (d\beta)^k$$

$$= -d(\alpha \wedge (d\alpha)^{q-k} \wedge (d\beta)^{k-1} \wedge \beta) + (d\alpha)^{q-k+1} \wedge (d\beta)^{k-1} \wedge \beta.$$

We claim that the second term is a sum of terms containing (q+1)-fold products of forms ω_j, and therefore vanishes. The form $d\alpha$ contains ω_j by (3.13), so each term of $(d\alpha)^{q-k+1}$ has products of ω_j's of length $q - k + 1$. Similarly $(d\beta)^{k-1}$ has in each term products of ω_j's of length $k - 1$ (using the integrability condition), while β itself is a linear combination of ω_j's. This proves the desired result.

Thus

$$\alpha' \wedge (d\alpha')^q - \alpha \wedge (d\alpha)^q = - d(\sum_{k=1}^{q} \begin{bmatrix} q \\ k \end{bmatrix} \alpha \wedge (d\alpha)^{q-k} \wedge d\beta)^k)$$

which completes the proof of Theorem 3.11. ∎

A case of particular interest is $n = 2q + 1$, when integration over the fundamental cycle yields a number, the Godbillon-Vey number of $\mathcal{F}$.

Generalizing Thurston's result for S^3 in [TH 1] mentioned in Chapter 2, Hurder proved that on odd-dimensional spheres S^{2q+1} admitting a rank q subbundle, every real number can be realized as the Godbillon-Vey number of a codimension q foliation [HU 6].

Interesting classes of foliations are obtained, as explained before, by the coset foliations of Lie groups by connected subgroups $H \subset G$. Let $\mathfrak{h} \subset \mathfrak{g}$ be the corresponding Lie algebras. Then the tangent bundle and normal bundle are given by $L = G \times \mathfrak{h}$ and $Q = G \times \mathfrak{g} / \mathfrak{h}$. If H is a closed connected subgroup, the coset foliation is defined by the submersion $G \to G/H$.

Let Γ be a discrete subgroup of G with orbit space $\Gamma \backslash G$ (left action). The foliation of G by H passes to a foliation of $\Gamma \backslash G$. These are locally homogeneous foliations. A typical example of such a foliation for $G = PSL(2,\mathbb{R})$ was discussed in Chapter 2.

Pullback of foliations. Let $f : M \to N$ be a smooth map, transverse to a foliation $\mathcal{F}$ on N. This means that $f_*(T_xM) + L_{f(x)} = T_{\mathcal{F}(x)}N$ for each $x \in M$. An example is a submersion and the foliation by points on N. For each leaf $\mathcal{L}$ of $\mathcal{F}$ on N consider $f^{-1}(\mathcal{L})$ and its connected components. Thus we obtain on M a foliation $f^*\mathcal{F}$ which has the same codimension as $\mathcal{F}$ on N.

In the next chapter we turn to a particularly interesting class of foliations.

CHAPTER 4
FLAT BUNDLES AND HOLONOMY

We begin by discussing a construction of <u>flat bundles</u>. Let $h : F \to F$ be a diffeomorphism of a smooth manifold F. The product foliation defined by the inverse images of the projection $\mathbb{R} \times F \to F$ is invariant under the action $(t,y)^n = (t+n, h^n(y))$ of h^n, $n \in \mathbb{Z}$. This means that the quotient $\mathbb{R} \times_{\mathbb{Z}} F$ carries a 1-dimensional foliation transverse to the fibers of $\mathbb{R} \times_{\mathbb{Z}} F \to \mathbb{R}/\mathbb{Z} \cong S^1$. The effect of h on a point x is obtained by the endpoint of the unique horizontal lift of S^1 with initial point x. Note that for the flow on $\mathbb{R} \times_{\mathbb{Z}} F$ transverse to the fibers, the map h is precisely the Poincaré map mentioned in the Introduction. The open Moebius band is of this type. For $F = (0,1)$ and $h(y) = -y$, it is the total space of the fibration $\mathbb{R} \times_{\mathbb{Z}} \mathbb{R} \to S^1$ with $(x,y) \sim (x+n, (-1)^n y)$ being the equivalence relation defining the total space. The leaves are circles, which are 2-fold coverings of the central circle $(y = 0)$, except for the central circle itself.

If $F = S^1$ and $h : S^1 \to S^1$ is the rotation through an angle α, then the resulting foliation on $\mathbb{R} \times_{\mathbb{Z}} S^1 = T^2$ is a linear foliation of the 2-torus as discussed in the Introduction.

This construction generalizes as follows. Let B be a manifold with fundamental group $\Gamma = \pi_1 B$. Consider a homomorphism $h : \Gamma \to \text{Diff } F$ into the group of diffeomorphisms of another manifold F. The group Γ acts on the universal covering space $\tilde{B}$ by deck transformations and on F via h, thus on the product $\tilde{B} \times F$ by

$$(\tilde{b}, y)^\gamma = (\tilde{b} \cdot \gamma, h(\gamma)y), \ \gamma \in \Gamma.$$

The projection $\tilde{B} \times F \to F$ defines a foliation $\tilde{\mathcal{F}}$, which is preserved by Γ. Thus the orbit space $\tilde{B} \times_\Gamma F$, defined by the equivalence relation $(\tilde{b}\cdot\gamma, y) \sim (\tilde{b}, h(\gamma^{-1})y)$, inherits a foliation $\mathcal{F}$, transverse to the fibers of the projection $\tilde{B} \times_\Gamma F \to \tilde{B}/\Gamma = B$.

The leaves carry locally the geometry of B, while the normal bundle of the foliation is the tangent bundle along the fibers of $\tilde{B} \times_\Gamma F \to B$. It carries any geometric structure of F preserved by the action of $h(\Gamma)$. For example, if Γ acts by isometries of a Riemannian structure on F, then a canonical Riemannian structure results on the normal bundle of the foliation. In this sense F models the normal or transverse geometry of the foliation, while B models its tangential or leaf geometry.

The representation of Γ by diffeomorphisms of F is called the <u>holonomy</u> of $\mathcal{F}$. It can be recovered from the structure of the flat bundle $\tilde{B} \times_\Gamma F \xrightarrow{\pi} B$ as follows. The restriction of the projection to each leaf is a covering map, and thus has the unique path lifting property. This defines for each path $\alpha : [0,1] \to B$ a unique diffeomorphism of fibers $\tau(\alpha) : \pi^{-1}(\alpha(0)) \to \pi^{-1}(\alpha(1))$, and thus for a closed path γ with basepoint b_0 a diffeomorphism $\tau(\gamma) : \pi^{-1}(b_0) \to \pi^{-1}(b_0)$ of the fiber over b_0. The covering homotopy property implies that $\tau(\gamma)$ depends only on the homotopy class of γ. Thus after identifying $\pi^{-1}(b_0)$ with F, the map $\tau(\gamma)$ can be identified with $h(\gamma)$.

PRINCIPAL BUNDLES. Let G be a Lie group with Lie algebra $\mathfrak{g}$, $P \to M$ a principal G-bundle on M, and ω a connection on P. Thus $\omega \in \Omega^1(P, \mathfrak{g})$, and its curvature is given by

$$\Omega = d\omega + \frac{1}{2}[\omega, \omega] \in \Omega^2(P, \mathfrak{g}).$$

Consider the horizontal subbundle $H \subset TP$, defined by $\omega = 0$.

4.1 LEMMA. H <u>is integrable if and only if</u> $\Omega = 0$.

PROOF. Note first that for arbitrary $X,Y \in \Gamma TP$

$$[\omega,\omega](X,Y) = [\omega(X),\omega(Y)] - [\omega(Y),\omega(X)] = 2[\omega(X),\omega(Y)].$$

Further for horizontal $X,Y \in \Gamma H$

$$d\omega(X,Y) = -\omega[X,Y].$$

If H is integrable, it follows that $\Omega = 0$. If conversely $\Omega = 0$, for horizontal X,Y we have then $d\omega(X,Y) = \Omega(X,Y) = 0$, and thus $\omega[X,Y] = 0$, which proves the horizontality of $[X,Y]$. Thus H is integrable. ∎

A connection ω with $\Omega = 0$ is a <u>flat connection</u>. A bundle $P \to M$ with a flat connection is <u>flat</u>. It carries a foliation transverse to the fibers. Its parallel transport defines the holonomy homomorphism $h : \Gamma = \pi_1 M \to G$ and

$$P \cong \tilde{M} \times_\Gamma G$$

where Γ acts by left translations on G. This is a special case of the situation described before.

The corresponding structure theorem for a smooth (not necessarily principal) fibration $F \to E \to B$ is as follows. Let $H \subset TE$ be an integrable subbundle which is transverse to the fiber at every point. If F is compact, then there is a homomorphism $h : \Gamma = \pi_1 B \to \text{Diff } F$ and a diffeomorphism

$$E \cong \tilde{B} \times_\Gamma F.$$

The interest of the flat bundle construction stems from the fact that locally every foliation has this structure. The idea is to take a tubular neighborhood of a leaf $\mathcal{L}$. It has the structure of a flat bundle as described above. This applies in particular to its restriction to a path α in a leaf $\mathcal{L}$.

This allows one to define the <u>holonomy for any foliation</u>. In the general case there is no longer a manifold F transverse to all leaves. A transversal manifold T is a submanifold of M such that at each point $x \in T$ the tangent space is a direct complement to $L_x \subset T_x M$. A transversal manifold may not hit each leaf, or may intersect a leaf in a complicated set. In the situation of a flat bundle described above, all fibers are transversal manifolds, and the intersection with each leaf is discrete. In the general case, there exist transversal manifolds through each point. The holonomy is then defined for each path α in a leaf $\mathcal{L}$ as a germ of diffeomorphisms $\tau(\alpha)$ on the germ of transversal manifolds at the initial point $\alpha(0)$ to the germ of transversal manifolds at $\alpha(1)$, obtained by sliding transversal manifolds along the path α (for a foliation defined by a submersion, the holonomy is trivial). In general, the map $\tau(\alpha)$ depends only on the homotopy class of the path α in the given leaf. For a loop γ in a leaf $\mathcal{L}$ with basepoint x_0, we have therefore a germ of diffeomorphisms $h(\gamma)$, mapping the germ of transversal manifolds at x_0 to itself. h is a representation of $\pi_1(\mathcal{L}, x_0)$ by diffeomorphisms on the germ of transversal manifolds at x_0.

This is the generalization of Poincaré's first return map from flows to foliations, and is an important tool in the theory of foliations.

As an illustration of these ideas, we consider the case of a transversally orientable foliation $\mathcal{F}$ of codimension one as in Chapter 2. It is defined by a nonsingular 1-form ω that is completely integrable. We consider a loop γ at x in the leaf $\mathcal{L}$ through $x \in M$. The tubular neighborhood of $\mathcal{L}$ restricted to the loop γ leads then to a 2-dimensional surface D, with two boundary components, of which one is γ. To describe the other, consider a point y of a transversal T through x. Then the path lift $\tilde{\gamma}$ of γ in the leaf through y leads to an endpoint $y' \in T$. The holonomy transformation $h(\gamma)$ is precisely given by $y' = h(\gamma)y$. The transversal path $\alpha_{y'y}$ in T completes $\tilde{\gamma}$ to a loop. The 2-dimensional region D is then bounded by $\tilde{\gamma} \cup \alpha \cup -\gamma$. Thus

$$(4.2) \qquad \int_{\alpha_{y'y}} \omega = \int_{\partial D} \omega = \int_{D} d\omega .$$

The first equality is based on the fact that $\int_{\gamma} \omega = 0$ and $\int_{\tilde{\gamma}} \omega = 0$, since both γ and $\tilde{\gamma}$ are paths in leaves of the foliation defined by ω. Note that for a tangent vector Z to T we have $\omega(Z) \neq 0$. It follows that the LHS is 0, if and only if $y' = y$ for all $y \in T$, i.e. $h(\gamma)$ = identity.

For the particular case of a foliation $\mathcal{F}$ defined by a closed 1-form ω, this implies the following fact due to Reeb [R1].

4.3 THEOREM. <u>Let $\mathcal{F}$ be a foliation of codimension one on M^n. If $\mathcal{F}$ is defined by a closed nonsingular 1-form, then the holonomy of every leaf is trivial.</u>

Another observation of Reeb [R1] is as follows.

4.4 PROPOSITION. <u>Let</u> M <u>be closed and</u> $\mathcal{F}$ <u>a foliation of codimension one defined by a closed nonsingular</u> 1-<u>form</u> ω. <u>Then there exists a transversal vector field, whose flow consists of diffeomorphisms preserving</u> $\mathcal{F}$, <u>i.e. mapping leaves into leaves</u>.

PROOF. Let Z be a vector field satisfying $\omega(Z) = 1$. Then

$$\theta(Z)\omega = i(Z)d\omega + di(Z)\omega = 0.$$

It follows that the flow φ_t of Z satisfies $\varphi_t^*\omega = \omega$, which proves the desired result. ■

Note that the flow φ_t maps a leaf $\mathcal{L}$ into a diffeomorphic leaf $\varphi_t(\mathcal{L})$. For a fixed $\mathcal{L}$ consider $U = \bigcup_{t\in\mathbb{R}} \varphi_t(\mathcal{L})$. Then $U \subset M$ is open and closed, hence coincides with M for connected M.

4.5 COROLLARY. <u>In the situation of Proposition</u> 4.4 <u>all leaves of</u> $\mathcal{F}$ <u>are diffeomorphic</u> (M <u>is assumed connected</u>).

Note that the map $\mathcal{L} \times \mathbb{R} \to M$ defined by $(x,t) \to \varphi_t(x)$ gives a product structure to a tubular neighborhood of $\mathcal{L}$ in M. This proves again Theorem 4.3 (but this argument involves the completeness of the flow φ_t).

For the linear foliations on the 2-torus Corollary 4.5 applies as well to the foliation with dense leaves (diffeomorphic to $\mathbb{R}$), as to those with rational slope (and circle leaves). The latter foliations are defined by submersions, but not the first type.

The following result of Tischler is of interest in this context.

4.6 THEOREM [TI]. Let M be a closed manifold with a closed nonsingular 1-form ω. Then there is a fibration $f : M \to S^1$. Let $\omega' = f^*d\theta$. Then the fibration can be chosen such that $\|\omega' - \omega\| < \epsilon$, where $\epsilon > 0$ is any prescribed number.

Here $\|\ \|$ denotes the global norm on forms, defined by a Riemannian metric on M. In terms of the foliations $\mathcal{F}, \mathcal{F}'$ defined by the closed 1-forms ω, ω', the statement is that for given $\mathcal{F}$ there exists arbitrarily close foliations $\mathcal{F}'$ which are given by fibrations $M \to S^1$.

Tischler's Theorem characterizes the closed manifolds fibering over S^1 as those admitting a foliation of codimension one of the type described. A particularly interesting situation is the one, when the typical fiber F itself carries a foliation preserved by a diffeomorphism $h : F \to F$. Then the construction at the beginning of this chapter yields a manifold $M \cong \mathbb{R} \times_{\mathbb{Z}} F$ fibering over the circle S^1 (and a further 1-dimensional foliation transverse to the fiber). For the case of a torus fiber $F = T^2$, the resulting 3-manifold has been analyzed in many special cases. An interesting example is the case of a matrix $A \in SL(2,\mathbb{Z})$, e.g. $A = \begin{bmatrix} 1 & 1 \\ 1 & 2 \end{bmatrix}$. Since A preserves the integral lattice $\mathbb{Z}^2$, it induces a diffeomorphism A_0 of the torus $T^2 = \mathbb{R}^2/\mathbb{Z}^2$. If, as in the example above, $\text{tr } A > 2$, the characteristic polynomial has two real eigenvalues, and the corresponding eigenvectors give rise to two complementary one-dimensional foliations invariant under the induced diffeomorphism A_0 of T^2. The resulting type of 3-manifolds T_A has been considered by Ghys-Sergiescu [GHS], Carrière [CA1,2] and also occurs prominently in Thurston's work [TH 6]. There it is shown to have a canonical partition into pieces which all have a simple geometric structure. This is a typical situation conjectured by Thurston to hold for all closed 3-manifolds.

PROOF OF THEOREM 4.6. We consider the universal covering $p : \tilde{M} \to M$ with basepoints $\tilde{x}_0$ and $x_0 = p(\tilde{x}_0)$. Then for $\tilde{x} \in \tilde{M}$ the integral $\int_{\tilde{x}_0}^{\tilde{x}} p^*\omega$ does not depend on the choice of the path γ from $\tilde{x}_0$ to $\tilde{x}$, since $d\omega = 0$. It follows that this integral defines a map $f_\omega : \tilde{M} \to \mathbb{R}$. An equivalent fact is that the map $\gamma \mapsto \int_\gamma \omega$ on the loops γ of M at x_0 depends only on the homotopy class of $\gamma \in \pi_1(M,x_0)$. Thus there is an induced map Per : $\pi_1(M,x_0) \to \mathbb{R}$ which is a homomorphism. Note that under the isomorphisms

$$\mathrm{Hom}(\pi_1 M, \mathbb{R}) \cong \mathrm{Hom}(H_1(M,\mathbb{R}),\mathbb{R}) \cong H^1_{DR}(M),$$

Per corresponds to the De Rham class $[\omega]$. The image $\mathrm{Per}(\omega) \subset \mathbb{R}$ is the group generated by the periods of ω. The case $[\omega] = 0$ is excluded by the compactness assumption on M, since a function g with $\omega = dg$ would give rise to singularities of ω at the critical points of g.

The map $f_\omega : \tilde{M} \to \mathbb{R}$ is equivariant with respect to Per, i.e.

$$f_\omega(\tilde{x} \cdot \gamma) = f_\omega(\tilde{x}) + \mathrm{Per}(\gamma),$$

and thus induces a map of quotients

$$f_\omega : M \to \mathbb{R}/\mathrm{Per}(\omega).$$

There are two possibilities. (i) The periods of ω are rationally related, and the group $\mathrm{Per}(\omega)$ is infinite cyclic. (ii) The periods of ω are not rationally related, and the group $\mathrm{Per}(\omega)$ is dense in $\mathbb{R}$.

In the first case, the map f_ω is a fibration $M \to S^1$. In the second case, we show that for $\epsilon > 0$ there is a closed nonsingular 1-form ω' with rationally related periods, and such that the global norm of $\omega' - \omega$ is $< \epsilon$.

Let $\omega = \omega_0 + dg$ be the De Rham-Hodge decomposition of ω with respect to a Riemannian metric. ω_0 denotes the harmonic representative of $[\omega]$ and g is a function. There is no term $\delta\alpha$, $\alpha \in \Omega^2(M)$, since $d\omega = d\delta\alpha = 0$, and thus $\langle d\delta\alpha, \alpha \rangle = \langle \delta\alpha, \delta\alpha \rangle = 0$ for the global scalar product, hence $\delta\alpha = 0$. The space of harmonic 1-forms $\mathcal{X}^1$ is isomorphic to $H^1_{DR}(M)$, in which the rational points $H^1(M,\mathbb{Q})$ are dense. These are represented by harmonic forms with rational periods. Thus for any $\epsilon > 0$ we can find $\omega'_0 \in \mathcal{X}^1$ with rational periods, and such that $\|\omega'_0 - \omega_0\| < \epsilon$. The form $\omega' = \omega'_0 + dg$ for sufficiently small ϵ remains nonsingular, and thus has the desired property. ∎

Note that a closed 1-form ω with rational periods yields an integer cohomology class $N \cdot [\omega] \in H^1(M,\mathbb{Z})$, after multiplying by an appropriate integer. Under the isomorphism

$$[M, K(\mathbb{Z},1)] \xrightarrow{\cong} H^1(M,\mathbb{Z}),$$

the class $N \cdot [\omega]$ corresponds to a map $M \to S^1 = K(\mathbb{Z},1)$ by pulling back the canonical generator (here $K(\mathbb{Z},1)$ denotes an Eilenberg-MacLane space with first homotopy group $\mathbb{Z}$, and all other homotopy groups trivial).

The modification of the given closed 1-form in the proof of Theorem 4.6 changes the cohomology class $[\omega]$, in case ω has irrational periods. This cannot be achieved by an isotopy of M, i.e. a diffeomorphism connected by a path to the identity of M.

This raises the following question: are two closed 1-forms ω_0, ω_1, defining the same cohomology class, related by an isotopy of M^2? Two 1-forms ω_0, ω_1 on M are isotopic, if there is a family φ_t of diffeomorphisms, $0 \leq t \leq 1$, with $\varphi_0 = $ id and $\omega_0 = \varphi_1^* \omega_1$. For two closed 1-forms ω_0, ω_1 a necessary condition is $[\omega_0] = [\omega_1]$, since $\varphi_t^* = $ id on cohomology. The following result is a partial answer, using the technique of Moser [MR].

4.7 THEOREM. Let M be a closed manifold, and ω_0, ω_1 nonsingular closed 1-forms with $[\omega_0] = [\omega_1] \in H^1_{DR}(M)$. If there is a family ω_t of nonsingular closed 1-forms coinciding with ω_0, ω_1 for $t = 0,1$ and such that $[\omega_t] \in H^1_{DR}(M)$ is independent of t, then ω_0, ω_1 are isotopic. This is in particular the case if there is a vector field (positively) transversal to both foliations defined by ω_0, ω_1.

PROOF. The given 1-forms can be connected by

$$\omega_t = (1-t)\omega_0 + t\omega_1, \quad 0 \leq t \leq 1.$$

Clearly $d\omega_t = 0$, since ω_0 and ω_1 are closed. To see that the cohomology class $[\omega_t]$ is independent of t, observe that

$$\omega_t - \omega_0 = t(\omega_1 - \omega_0).$$

Thus for f with $\omega_1 - \omega_0 = df$ we have

$$\omega_t - \omega_0 = d(tf)$$

and $[\omega_t] = [\omega_0]$. If there is a vector field Z such that both $\omega_0(Z) > 0$ and $\omega_1(Z) > 0$, then $\omega_t(Z) > 0$ and ω_t is nonsingular for each t. This is the condition of positive transversality in the statement of the Theorem. In general it may not be possible to connect ω_0, ω_1 by cohomologous and nonsingular 1-forms ω_t. This is in fact an extremely difficult question (for a positive answer in dimension three, and its relation to Cerf's Theorem, see [LB]). That is why the existence of ω_t as in the Theorem has to be postulated.

Given a family ω_t, the problem is to find a family of diffeomorphisms φ_t, $0 \leq t \leq 1$, $\varphi_0 = \mathrm{id}$, such that

(4.8) $$\varphi_t^* \omega_t = \text{constant} = \omega_0.$$

A necessary condition is obtained by differentiating this identity with respect to t. Let X_t be the (time-dependent) vector field generating the family φ_t (X_t is autonomous precisely for a 1-parameter group φ_t of diffeomorphisms, a condition not involved in our discussion). We obtain (see e.g. [GS, p. 110]) with $\cdot = \frac{\partial}{\partial t}$

(4.9) $$(\varphi_t^* \omega_t)^{\cdot} = \varphi_t^* \theta(X_t) \omega_t + \varphi_t^* \dot{\omega}_t = 0.$$

Since by assumption $d\omega_t = 0$, this yields

(4.10) $$\varphi_t^* di(X_t)\omega_t + \varphi_t^* \dot{\omega}_t = 0.$$

Now we can use the cohomology assumption, which implies

$$\omega_t = \omega_0 + dG_t \quad \text{with} \quad G_t : M \to \mathbb{R}.$$

Hence

$$\dot{\omega}_t = dg_t \quad \text{with} \quad g_t = \dot{G}_t.$$

It follows from (4.10) that

$$\varphi_t^* d(i(X_t)\omega_t + g_t) = 0.$$

Since φ_t is a diffeomorphism, this is equivalent to

$$d(i(X_t)\omega_t + g_t) = 0.$$

A sufficient condition for X_t is therefore

(4.11) $$i(X_t)\omega_t + g_t = 0.$$

Let Z_t be a transversal vectorfield to the foliation $\mathcal{F}_t$ defined by ω_t, and satisfying $i(Z_t)\omega_t = 1$ (here is the place where we use the nonsingularity assumption on ω_t). Defining

$$X_t = - g_t Z_t$$

yields a solution to (4.11). Integrating the vector field X_t yields then an isotopy φ_t satisfying (4.8). ∎

CHAPTER 5

RIEMANNIAN AND TOTALLY GEODESIC FOLIATIONS

The <u>transversal geometry</u> of a foliation is the geometry infinitesimally modeled by Q, while the tangential geometry is infinitesimally modeled by L. A key fact is the existence of the Bott connection in Q defined by

(5.1) $\overset{\circ}{\nabla}_X s = \pi[X, Y_s]$ for $X \in \Gamma L$, $s \in \Gamma Q$

where $Y_s \in \Gamma TM$ is any vectorfield projecting to s under $\pi : TM \to Q$. It is a partial connection along L (only defined for $X \in \Gamma L$), but otherwise satisfies the usual connection properties. First we observe that the RHS in (5.1) is independent of the choice of Y_s. Namely the difference of two such choices is a vector field $X' \in \Gamma L$, and $[X, X'] \in \Gamma L$ so that $\pi[X, X'] = 0$.

The curvature $\overset{\circ}{R}(X,Y) = \overset{\circ}{\nabla}_X \overset{\circ}{\nabla}_Y - \overset{\circ}{\nabla}_Y \overset{\circ}{\nabla}_X - \overset{\circ}{\nabla}_{[X,Y]}$ for $X, Y \in \Gamma L$ is zero, as a consequence of the Jacobi identity for the bracket of vector fields. This means that Q restricted to each leaf $\mathcal{L}$ is a flat vector bundle. The parallel transport in Q along a path in $\mathcal{L}$ is the linearized holonomy discussed before. Here Q plays the role of the tangent bundle to the (germs of) transversal manifolds of $\mathcal{F}$. The vanishing of $\overset{\circ}{R}$ is equivalent to the property, that the parallel transport in Q depends only on the homotopy class of a path in a leaf.

An adapted connection in Q is a connection restricting along L to the partial connection $\overset{\circ}{\nabla}$ given by (5.1). To show that such connections exist, consider a Riemannian metric g_M on M. Then TM splits orthogonally as

(5.2) $TM = L \oplus L^\perp$

with $\sigma : Q \xrightarrow{\cong} L^\perp \subset TM$ splitting the sequence (3.4). The metric g_M on TM is then a direct sum

$$g_M = g_L \oplus g_{L^\perp}.$$

With $g_Q = \sigma^* g_{L^\perp}$, the splitting map $\sigma : (Q,g_Q) \to (L^\perp, g_{L^\perp})$ is a metric isomorphism. Let now ∇^M be the Levi-Cività connection associated to the Riemannian metric g_M on M. Then for $s \in \Gamma Q$ and $Z_s = \sigma(s) \in \Gamma L^\perp$ the definition

(5.3) $\quad \nabla_X s = \begin{cases} \pi[X, Z_s] & \text{for } X \in \Gamma L, \\ \pi(\nabla^M_X Z_s) & \text{for } X \in \Gamma L^\perp, \end{cases}$

yields an adapted connection ∇ in Q. Its curvature R_∇ coincides with $\overset{\circ}{R}$ for $X, X' \in \Gamma L$, hence $R_\nabla(X, X') = 0$ for $X, X' \in \Gamma L$.

A connection ∇ in Q defines a connection ∇^* in Q^* by the formula

$$(\nabla^*_X \omega)(s) = X\omega(s) - \omega(\nabla_X s)$$

for $X \in \Gamma TM$, $\omega \in \Gamma Q^*$ and $s \in \Gamma Q$. For the partial Bott connection $\overset{\circ}{\nabla}$ given by (5.1), this yields the formula

$$(\overset{\circ}{\nabla}^*_X \omega)(Y) = X\omega(Y) - \omega[X, Y]$$

for $X \in \Gamma L$ and $Y \in \Gamma TM$ with $\pi(Y) = s$. Thus

(5.4) $\quad \overset{\circ}{\nabla}^*_X \omega = \theta(X)\omega \quad \text{for} \quad \omega \in \Gamma^* Q \subset \Omega^1(M).$

For any connection ∇ in Q there is a torsion $T_\nabla \in \Omega^2(M,Q)$ defined by

$$T_\nabla(Y,Y') = \nabla_Y \pi(Y') - \nabla_{Y'} \pi(Y) - \pi[Y,Y']$$

for $Y,Y' \in \Gamma TM$.

5.5 PROPOSITION. <u>For any metric</u> g_M <u>on</u> M, <u>and the connection</u> ∇ <u>on</u> Q <u>defined by</u> (5.3), <u>we have</u> $T_\nabla = 0$.

PROOF. For $X \in \Gamma L$, $Y \in \Gamma TM$ we have $\pi(X) = 0$ and

$$T_\nabla(X,Y) = \nabla_X \pi(Y) - \pi[X,Y] = 0$$

by (5.1). For $Z,Z' \in \Gamma L^\perp$ we have

$$T_\nabla(Z,Z') = \pi(\nabla^M_Z Z') - \pi(\nabla^M_{Z'} Z) - \pi[Z,Z'] = \pi(T_{\nabla^M}(Z,Z')) = 0,$$

where T_{∇^M} is the (vanishing) torsion ∇^M. Finally the bilinearity and skew-symmetry of T_∇ gives the desired result. ∎

At this point it will be helpful to explain the general concept of a G-foliation. Let G be a Lie subgroup of $GL(q,\mathbb{R})$. Consider an atlas $\mathcal{U} = \{U_\alpha\}$ of distinguished charts, $f_\alpha : U_\alpha \to \mathbb{R}^q$ submersions defining $\mathcal{F}|U_\alpha$, related by transition functions $\gamma_{\beta\alpha} : U_\alpha \cap U_\beta \to GL(q,\mathbb{R})$ as in (3.1). $\mathcal{F}$ is a G-foliation, if the atlas $\mathcal{U}$ can be chosen such that for all α,β with $U_\alpha \cap U_\beta \neq \phi$ the derivatives of the transition functions define maps

$$(\gamma_{\alpha\beta})_* : U_\alpha \cap U_\beta \to G \subset GL(q,\mathbb{R}).$$

For the foliation of M by points this is the usual concept of a G-structure. We refer to [D][KT3,4] for many examples of G-foliations.

For the case $G = SL(q)$ this means the following. The foliation has to be transversally orientable, i.e. there must exist a nowhere vanishing section $\nu \in \Lambda^q \mathbf{Q}^*$, a transversal volume form, such that

(5.6) $\theta(X)\nu = 0$ for all $X \in \Gamma L$.

This condition is called the holonomy invariance of ν. Here the LHS is defined by

$$(\theta(X)\nu)(s_1,\ldots,s_q) = X\nu(s_1,\ldots,s_q) - \sum_{i=1}^{q} \nu(s_1,\ldots,\nabla_X s_i,\ldots,s_q)$$

for $s_1,\ldots,s_q \in \Gamma\mathbf{Q}$. Since $X \in \Gamma L$, the RHS only involves the canonical Bott connection $\overset{\circ}{\nabla}$, and not its extension to a full-fledged connection via g_M. Observe that $i(X)\nu = 0$ for $X \in \Gamma L$. Thus

$$\theta(X)\nu = i(X)d\nu.$$

If α is a 1-form satisfying $d\nu = \alpha \wedge \nu$ as in (3.10), it follows that

$$\theta(X)\nu = i(X)(\alpha \wedge \nu) = \alpha(X) \cdot \nu.$$

As a consequence, for a SL(q)-foliation $\mathcal{F}$ we find that the Godbillon-Vey class $[\alpha \wedge (d\alpha)^q] = 0$. This class can be viewed as an obstruction to the existence of a SL(q)-structure for $\mathcal{F}$.

To further interpret the formula $\theta(X)\nu = \alpha(X) \cdot \nu$, observe that by (5.4) this means that α is the connection form for the canonical induced connection in the line bundle $\Lambda^q Q^*$. The condition $\theta(X)\nu = 0$ for all $X \in \Gamma L$ says that γ is a parallel section in $\Lambda^q Q^*$, i.e. invariant under the holonomy maps of Q, associated to paths in the leaves of $\mathcal{F}$ with respect to the canonical connection.

For the purpose of this text, the most important case is the case of a <u>Riemannian foliation</u> with $G = O(q)$, or $SO(q)$ in the case of a <u>transversally oriented</u> Riemannian foliation. The study of these foliations was initiated by Reinhart [RE 2]. The requirement is that the local transition functions $\gamma_{\beta\alpha}$ are isometries of suitably given Riemannian metrics on $\mathbb{R}^q$. (Note that these need not be the Euclidean metric, but just any Riemannian metrics, with possibly nontrivial curvature.). The local submersions f_α define then by pull-back a Riemannian metric g_Q on the normal bundle Q, invariantly defined because of the isometric property of the $\gamma_{\beta\alpha}$. For this metric it follows then that

(5.7) $\theta(X)g_Q = 0$ for all $X \in \Gamma L$.

This condition is called the holonomy invariance of g_Q. It is the infinitesimal equivalent of the invariance under the holonomy transformations sketched in Chapter 4 on transversal manifolds, and serves as the technical definition of the Riemannian property. A metric g_M on M is <u>bundle-like</u>, if the induced metric g_Q on Q is holonomy invariant.

The simplest example is given by a nonsingular Killing vector field X on (M, g_M). This means that $\theta(X)g_M = 0$ or equivalently

$$Xg_M(Y, Y') = g_M([X, Y], Y') + g_M(Y, [X, Y'])$$

for any vector fields $Y, Y' \in \Gamma TM$. Let $\mathcal{F}$ be the foliation of M by the orbits of X. Then X is a nontrivial section of $L \subset TM$. The complement $L^\perp$ is preserved by the flow and for the induced metric g_Q we have $\theta(X)g_Q = 0$. The holonomy invariance in this case is precisely the invariance under the flow generated by X.

More generally consider a Lie group G, acting by isometries on (M, g_M). If the orbits of the G-action have all the same dimension, this gives rise to a Riemannian foliation. The point is that the sections of L are linear combinations of Killing vector fields arising from the group action, so that the previous arguments apply. This situation occurs in particular for actions of compact groups, since any metric on M can be averaged to an invariant metric under the action.

We return now to the general situation of a Riemannian foliation on (M, g_M). The Lie derivative for any metric g_Q on Q is given by

$$(5.8) \qquad (\theta(X)g_Q)(s,t) = Xg_Q(s,t) - g_Q(\nabla_X s, t) - g_Q(s, \nabla_X t)$$

for $X \in \Gamma L$ and $s, t \in \Gamma Q$. Again the RHS involves only the canonical Bott connection $\overset{\circ}{\nabla}$ (and not its extension to an adapted connection ∇). More generally for any covariant r-tensor ω on Q we have

$$(\theta(X)\omega)(s_1, \ldots, s_r) = X\omega(s_1, \ldots, s_r) - \sum_{i=1}^{r} \omega(s_1, \ldots, \nabla_X s_i, \ldots, s_r)$$

for $X \in \Gamma L$ and $s_1, \ldots, s_r \in \Gamma Q$. Condition (5.7) is therefore equivalent to the identity

$$(5.9) \qquad Xg_Q(s,t) = g_Q(\pi[X, Z_s], t) + g_Q(s, \pi[X, Z_t])$$

for $X \in \Gamma L$, sections $s,t \in \Gamma Q$ and $Z_s = \sigma(s)$, $Z_t = \sigma(t) \in \Gamma L^\perp$. It is interesting to compare this with the condition that the connection ∇ defined by (5.3) is a metric connection in the bundle Q equipped with the induced metric g_Q. This condition reads for $s,t \in \Gamma Q$

(5.10) $\quad Y g_Q(s,t) = g_Q(\nabla_Y s, t) + g_Q(s, \nabla_Y t)$

but now for all $Y \in \Gamma TM$ (not only $X \in \Gamma L$), and thus implies (5.9).

5.11 THEOREM. Let $\mathcal{F}$ be a foliation on (M, g_M), g_Q the induced metric on Q, and ∇ the connection on Q defined by (5.3). Then $\mathcal{F}$ is Riemannian and g_M bundle-like, if and only if ∇ is a metric connection.

PROOF. It suffices to verify that for $\mathcal{F}$ Riemannian the condition (5.10) holds for g_Q and ∇ as in the Theorem. It suffices to verify this for $Z \in \Gamma L^\perp$. But then we have for $Z_s = \sigma(s)$, $Z_t = \sigma(t)$

$$Z g_Q(s,t) = Z g_M(Z_s, Z_t)$$

$$= g_M(\nabla^M_Z Z_s, Z_t) + g_M(Z_s, \nabla^M_Z Z_t)$$

$$= g_Q(\pi(\nabla^M_Z Z_s), t) + g_Q(s, \pi(\nabla^M_Z Z_t))$$

$$= g_Q(\nabla_Z s, t) + g_Q(s, \nabla_Z t). \blacksquare$$

5.12 THEOREM. Let g_Q be a holonomy invariant metric in the normal bundle Q of $\mathcal{F}$. Then there is a unique metric and torsion-free connection in Q.

PROOF. The existence follows by constructing ∇ via a bundle-like metric g_M. It remains to prove the uniqueness. Let ∇ be a metric and torsion-free connection in Q. Then

$$
\begin{aligned}
2g_Q(\nabla_Y s,t) = {} & Yg_Q(s,t) + Z_s g_Q(\pi(Y),t) - Z_t g_Q(\pi(Y),s) \\
& + g_Q(\pi[Y,Z_s],t) + g_Q(\pi[Z_t,Y],s) - g_Q(\pi[Z_s,Z_t],\pi(Y))
\end{aligned}
\tag{5.13}
$$

for $Y \in \Gamma TM$; $s,t \in \Gamma Q$; $Z_s, Z_t \in \Gamma TM$ with $\pi(Z_s) = s$, $\pi(Z_t) = t$. This formula is proved by expanding the first three terms on the RHS using (5.10), and then using torsion-freeness. (5.13) implies the uniqueness of ∇. ∎

The unique metric and torsion-free connection ∇ in the normal bundle of a Riemannian foliation $\mathcal{F}$ is the transversal Levi-Cività connection of $\mathcal{F}$. It is worth repeating that the (holonomy invariant) transversal metric g_Q determines ∇. Any bundle-like metric g_M inducing g_Q on Q leads to the same ∇. Formula (5.3) shows that the covariant derivative in the transversal directions corresponds under the local Riemannian submersion, to the effect of the Levi-Cività connections on the Riemannian manifolds modeling the foliation. The transition functions being isometries, the pull-backs are invariantly defined. In particular all curvature data associated to ∇ have an invariant meaning.

An additional important property of the curvature R_∇ of ∇ is

(5.14) $i(X)R_\nabla = 0$ for $X \in \Gamma L$.

We return to the situation of an arbitrary foliation on (M, g_M), and the metric g_Q induced on Q. Identifying $(Q, g_Q) \cong (L^\perp, g_M | L^\perp)$ we have then for $X \in \Gamma L$ and $Z, Z' \in \Gamma L^\perp$

(5.15) $\quad (\theta(X) g_Q)(Z, Z') = X g_Q(Z, Z') - g_Q(\pi[X,Z], Z') - g_Q(Z, \pi[X, Z'])$
$\quad\quad\quad\quad\quad\quad\quad = X g_M(Z, Z') - g_M([X,Z], Z') - g_M(Z, [X, Z']).$

Note that the vanishing of this bilinear form follows already from the vanishing of the corresponding quadratic form on unit vectors, hence $\mathcal{F}$ is Riemannian and g_M bundle-like, if and only if

(5.16) $\quad g_M([X,Z], Z) = 0,$

for all $X \in \Gamma L$ and $Z \in \Gamma L^\perp$ with $|Z| = 1$. For the torsionfree connection ∇^M we have

$$[X, Z] = \nabla^M_X Z - \nabla^M_Z X, \quad [X, Z'] = \nabla^M_X Z' - \nabla^M_{Z'} X.$$

Thus (5.15) can be rewritten as

(5.17) $\quad (\theta(X) g_Q)(Z, Z') = g_M(\nabla^M_Z X, Z') + g_M(Z, \nabla^M_{Z'} X).$

The following formula is useful

(5.18) $\quad (\theta(X) g_Q)(Z, Z') = - g_M(\nabla^M_Z Z' + \nabla^M_{Z'} Z, X)$

$\quad\quad\quad\quad\quad\quad\quad = - 2 g_M(\nabla^M_Z Z', X) + g_M([Z, Z'], X),$

for $X \in \Gamma L$ and $Z, Z' \in \Gamma Q$.

PROOF. We use

$$g_M(\nabla^M_Z X, Z') = Z g_M(X, Z') - g_M(X, \nabla^M_Z Z') = - g_M(X, \nabla^M_Z Z'),$$

and similarly

$$g_M(Z, \nabla^M_{Z'} X) = - g_M(\nabla^M_{Z'} Z, X).$$

By (5.17) we find

$$(\theta(X) g_Q)(Z, Z') = - g_M(\nabla^M_Z Z', X) - g_M(\nabla^M_{Z'} Z, X)$$

$$= - 2 g_M(\nabla^M_Z Z', X) + g_M([Z, Z'], X)$$

as claimed. ■

These formulas establish the following facts due to Reinhart.

5.19 THEOREM. Let $\mathcal{F}$ be a foliation on (M, g_M). Then the following conditions are equivalent:
(i) $\mathcal{F}$ is Riemannian and g_M bundle-like;
(ii) $g_M(\nabla^M_Z X, Z') + g_M(Z, \nabla^M_{Z'} X) = 0$ for $X \in \Gamma L$ and $Z, Z' \in \Gamma L^\perp$;
(iii) $g_M([X, Z], Z) = 0$ for $X \in \Gamma L$, $Z \in \Gamma L^\perp$ with $|Z| = 1$;
(iv) $g_M(\nabla^M_Z Z' + \nabla^M_{Z'} Z, X) = 0$ for $X \in \Gamma L$ and $Z, Z' \in \Gamma L^\perp$;
(v) $2 g_M(\nabla^M_Z Z', X) = g_M([Z, Z'], X)$ for $X \in \Gamma L$ and $Z, Z' \in \Gamma L^\perp$.

PROOF. (i)⇔(ii) follows from (5.17). (i)⇔(iii) was already explained (see (5.16)). (i)⇔(iv)⇔(v) follows from (5.18). ∎

It is of interest to consider the conditions obtained by switching the roles of L and $L^\perp$. Thus let again $\mathcal{F}$ be a foliation on (M, g_M), with induced metric g_L on L. For $Z \in \Gamma L^\perp$ and $X, X' \in \Gamma L$ we define then formally in analogy to (5.15)

$$(5.20) \quad (\theta(Z)g_L)(X,X') = Zg_L(X,X') - g_L(\pi^\perp[Z,X],X') - g_L(X,\pi^\perp[Z,X'])$$

$$= Zg_M(X,X') - g_M([Z,X],X') - g_M(X,[Z,X']).$$

The vanishing of this bilinear form follows from the vanishing of the corresponding quadratic form on unit vectors (polarization), hence from

$$(5.21) \quad g_M([Z,X],X) = 0,$$

for all $X \in \Gamma L$ with $|X| = 1$ and $Z \in \Gamma L^\perp$. Further by the calculation leading to (5.18) we find the formula

$$(\theta(Z)g_L)(X,X') = -2g_M(\nabla^M_X X', Z) + g_M([X,X'],Z).$$

But now X, X' are sections of the involutive bundle L, so that the second term on the RHS vanishes. It follows that

$$(5.22) \quad (\theta(Z)g_L)(X,X') = -2g_M(\nabla^M_X X', Z).$$

The vanishing of this expression for all $Z \in \Gamma L^\perp$ is equivalent to $\nabla^M_X X' \in \Gamma L$ for $X, X' \in \Gamma L$. This property is equivalent to the property that all leaves of $\mathcal{F}$ are totally geodesic submanifolds of (M,g_M) (see [KN]). A foliation satisfying these conditions is called totally geodesic. The following statement is a summary of results due to Cairns [C 1,2,3], Carrière-Ghys [CAGH] and Rummler [RU 1].

5.23 THEOREM. Let $\mathcal{F}$ be a foliation on (M,g_M). Then the following conditions are equivalent:

(i) $\mathcal{F}$ is totally geodesic, i.e. $g_M(\nabla_X^M X', Z) = 0$ for $X, X' \in \Gamma L$ and $Z \in \Gamma L^\perp$;

(ii) $\theta(Z) g_L = 0$ for $Z \in \Gamma L^\perp$;

(iii) $g_M(\nabla_X^M Z, X') + g_M(X, \nabla_X^M Z) = 0$ for $X, X' \in \Gamma L$ and $Z \in \Gamma L^\perp$;

(iv) $g_M([Z,X], X) = 0$ for $X \in \Gamma L$ with $|X| = 1$ and $Z \in \Gamma L^\perp$;

(v) $g_M(\nabla_Z^M X - [Z,X], X') = 0$ for $X, X' \in \Gamma L$ and $Z \in \Gamma L^\perp$.

PROOF. (i)$\Leftrightarrow$(ii) follows from (5.22). (ii)$\Leftrightarrow$(iii) follows from the calculation

$$(\theta(Z)g_L)(X,X') = Zg_M(X,X') - g_M(\nabla_Z^M X - \nabla_X^M Z, X') - g_M(X, \nabla_Z^M X' - \nabla_{X'}^M Z)$$
$$= g_M(\nabla_X^M Z, X') + g_M(X, \nabla_{X'}^M Z).$$

(ii)$\Leftrightarrow$(iv) has been explained before (see (5.21)). It remains to establish (i)$\Leftrightarrow$(v). We consider

$$g_M([Z,X], X') = g_M(\nabla_Z^M X - \nabla_X^M Z, X')$$
$$= g_M(\nabla_Z^M X, X') - (Xg_M(Z,X') - g_M(Z, \nabla_X^M X')).$$

Thus

$$g_M(\nabla^M_Z X, X') - g_M([Z,X], X') = - g_M(Z, \nabla^M_X X')$$

(v) is equivalent to the vanishing of the LHS, while (i) is equivalent to the vanishing of the RHS. ∎

As a consequence of (v) we find for a totally geodesic foliation the formula

(5.24) $\pi^\perp(\nabla^M_Z X) = \pi^\perp([Z,X])$ for $X \in \Gamma L$, $Z \in \Gamma L^\perp$.

Cairns [C 3] defines a connection ∇^L in L by

(5.25) $\nabla^L_Y X = \pi^\perp(\nabla^M_Y X)$ for any $Y \in \Gamma TM$, $X \in \Gamma L$.

It follows then from (5.24) that

(5.26) $\nabla^L_Z X = \pi^\perp([Z,X])$ for $Z \in \Gamma L^\perp$, $X \in \Gamma L$.

The analogy with the Bott connection (5.1) is clear. The difference is even more important: ∇^L depends on a metric g_M, and the resulting choice of $L^\perp$. Moreover (5.26) is true only because $\mathcal{F}$ is assumed totally geodesic. Cairns calls the connection ∇^L the tangential Levi-Civitá connection of $\mathcal{F}$. Restricted to a leaf $\mathcal{L} \subset M$ it is by (5.24) the usual induced connection for the submanifold $\mathcal{L}$ (the normal component representing the second fundamental form of $\mathcal{L} \subset M$). Note that its torsion $T_{\nabla^L} \in \Omega^2(M,L)$, given by

(5.27) $T_{\nabla^L}(Y,Y') = \nabla^L_Y \pi^\perp(Y') - \nabla^L_{Y'} \pi^\perp(Y) - \pi^\perp[Y,Y']$

for $Y,Y' \in \Gamma TM$, does not vanish. In fact for $Z,Z' \in \Gamma L^\perp$ one has

(5.28) $\quad T_{\nabla^L}(Z,Z') = - \pi^\perp[Z,Z']$.

For the case of a codimension one foliation, the connection ∇^L is thus torsion-free. (5.26) allows to rewrite condition (ii) in Theorem 5.23 as

$$Zg_L(X,X') = g_L(\nabla_Z^L X, X') + g_L(X, \nabla_Z^L X'),$$

i.e. ∇^L is a metric connection in (L, g_L).

The previous considerations are particularly clear for the case of a foliation $\mathcal{F}$ with integrable orthogonal bundle $L^\perp$. We are then in the presence of two orthogonal foliations $\mathcal{F}$ and $\mathcal{F}^\perp$ on (M, g_M). In this situation $\mathcal{F}$ is Riemannian if and only if $\mathcal{F}^\perp$ is totally geodesic ([JW 2] [CAGH]). Assuming this to be the case, (5.25)(5.26) show that the tangential Levi-Cività connection ∇^L of $\mathcal{F}$ is precisely the transversal Levi-Cività connection in the normal bundle of $\mathcal{F}^\perp$.

An illustrative example of such a situation is a flat bundle $M = \tilde{B} \times_\Gamma F \to B$ associated to a homomorphism $h : \Gamma = \pi_1 B \to \text{Diff } F$ as explained in Chapter 4. The foliation $\mathcal{F}$ by the fibers and the foliation $\mathcal{F}^\perp$ transversal to the fibers are orthogonal with respect to a metric g_M, induced by the product metric $p^* g_B \oplus g_F$ on $\tilde{B} \times F$. Here g_B denotes a metric on B and $p^* g_B$ its pull-back to the universal covering $\tilde{B}$, while g_F is a metric on F. Since the holonomy of $\mathcal{F}$ is clearly trivial, g_M is bundle-like for $\mathcal{F}$. Note that the induced metric in the normal bundle is g_B. As a consequence of our previous discussion, $\mathcal{F}$ is Riemannian and equivalently $\mathcal{F}^\perp$ totally geodesic. But note that $\mathcal{F}^\perp$ is not necessarily Riemannian. In fact, its normal bundle is the tangent bundle $T(f)$ along the fibers of $f : M \to B$, and its holonomy maps in $T(f)$ need not be isometries. For the diffeomorphism $T^2 \to T^2$ induced by $A = \begin{bmatrix} 1 & 1 \\ 1 & 2 \end{bmatrix} \in SL(2,\mathbb{Z})$ e.g., the foliation transverse to the fibers of the resulting torus fibration $T_A = \mathbb{R} \times_{\mathbb{Z}} T^2 \to S^1$ has nonisometric holonomy, hence cannot be Riemannian. Another example is the Roussarie foliation on the unit tangent bundle of a Riemannian surface M_g ($g > 1$), considered in Chapter 2. It is a codimension one foliation transverse to the circle fibers in $T_1 M_g \cong D^2 \times_\Gamma S^1$, $\Gamma = \pi_1(M_g)$. This foliation is definitely not Riemannian. This follows from the nontriviality of its Godbillon-Vey class. Namely a (transversally oriented) Riemannian foliation is in particular a $SL(q)$-foliation, and therefore its Godbillon-Vey class vanishes, by a remark made early in this chapter.

If however $\tilde{B} \times_\Gamma F \to B$ arises from a representation of Γ by isometries of a Riemannian metric g_F on F, then this will turn $\mathcal{F}^\perp$ into a Riemannian foliation, and hence $\mathcal{F}$ into a totally geodesic foliation.

For foliations $\mathcal{F}$ of codimension one the situation is particularly simple, since the complementary foliation is necessarily integrable, namely a flow. It follows that $\mathcal{F}$ is totally geodesic, iff the transversal flow is Riemannian, an observation already made by Reinhart [RE 2] and Rummler [RU 1]. This fact underlies Carrière's classification theorem [CA 1][CA 2] for Riemannian flows on closed 3-manifolds, and the classification by Carrière-Ghys [CAGH] and Ghys [GH 1] of totally geodesic foliations of codimension one on closed 3-manifolds.

CHAPTER 6

SECOND FUNDAMENTAL FORM AND MEAN CURVATURE

Let (M, g_M) be a Riemannian manifold. For a submanifold $\mathcal{L} \subset M$, and vector fields X, X' tangent to $\mathcal{L}$, the second fundamental form $\alpha(X, X')$ takes values in the normal bundle, and is given by

$$(6.1) \qquad \alpha(X, X') = \pi(\nabla_X^M X'),$$

where π is the projection onto the normal bundle. For a foliation $\mathcal{F}$ on (M, g_M) this formula yields a bundle map $\alpha : L \otimes L \to Q$. The involutivity of L shows that α is symmetric. In fact the definition $\alpha = -\nabla \pi$ for $\pi \in \Omega^1(M, Q)$ yields even a more general symmetric form $TM \otimes TM \to Q$, that restricts to the α above (see [KT6, p. 94]). But here we use α in the restricted sense (6.1). Note that for $Z \in \Gamma L^\perp$

$$(6.2) \qquad g_Q(\alpha(X, X'), Z) = g_M(\nabla_X^M X', Z) = -g_M(X', \nabla_X^M Z).$$

From this we conclude that $\mathcal{F}$ is totally geodesic exactly when $\alpha = 0$.

If we associate to α a linear (Weingarten) map $W(Z) : L \to L$ for $Z \in \Gamma L^\perp$, by the formula

$$(6.3) \qquad g_Q(\alpha(X, X'), Z) = g_L(W(Z)X, X'),$$

then $W(Z)$ is self-adjoint. We find from (6.2) the usual formula

$$(6.4) \qquad W(Z)X = -\pi^\perp(\nabla_X^M Z).$$

An interesting interpretation of α is obtained from (5.22) and (6.2):

(6.5) $\quad (\theta(Z)g_L)(X,X') = -2g_Q(\alpha(X,X'),Z)$,

i.e. for an orthogonal vectorfield Z to $\mathcal{F}$ the Z-component of α is the Lie derivative with respect to Z of the metric along the leaves.

6.6 COROLLARY. Let $\mathcal{F}$ be a foliation on (M,g_M). Then $\mathcal{F}$ is a totally geodesic foliation, iff the induced metric g_L along the leaves is invariant under flows of vectorfields orthogonal to the foliation.

Using the symmetry of α, we note that $\alpha : L \otimes L \to Q$ is equivalently given by

(6.7) $\quad \alpha(X,X') = \frac{1}{2} \pi(\nabla_X^M X' + \nabla_{X'}^M X)$.

Now we reverse the roles of L and $L^\perp$. Modifying slightly an idea of Reinhart [RE 7,8], and motivated by the last formula, we define the second fundamental form $\alpha^\perp$ of $L^\perp$ by

(6.8) $\quad \alpha^\perp(Z,Z') = \frac{1}{2} \pi^\perp(\nabla_Z^M Z' + \nabla_{Z'}^M Z)$

for $Z,Z' \in \Gamma L^\perp$. Then $\alpha^\perp : L^\perp \otimes L^\perp \to L$ is a symmetric bundle map. By (5.18) we find the formula

(6.9) $\quad (\theta(X)g_Q)(Z,Z') = -2g_L(\alpha^\perp(Z,Z'),X)$.

6.10 COROLLARY. Let $\mathcal{F}$ be a foliation on (M, g_M). Then $\mathcal{F}$ is Riemannian and g_M bundle-like iff $\alpha^\perp = 0$.

Since a Riemannian foliation is locally given by Riemannian submersions, locally a curve everywhere tangent to $L^\perp$ projects onto a curve of the same length, hence a curve projecting onto a geodesic is itself a geodesic. It follows that a geodesic which is tangent to $L^\perp$ at one point, remains tangent to $L^\perp$ at each of its points. This is the total geodesic property of $L^\perp$. By the above $\alpha^\perp = 0$ implies this property. Conversely the total geodesic property of $L^\perp$ implies $\alpha^\perp = 0$. To see this, extend a unit normal vector to a unit vector field $Z \in \Gamma L^\perp$, and tangent to a geodesic at each point. Then $g_M(\nabla^M_Z Z, X) = 0$ for all $X \in \Gamma L$. It follows that $\alpha^\perp(Z,Z) = 0$ for $Z \in \Gamma L^\perp$. Since $\alpha^\perp$ is symmetric, this implies $\alpha^\perp = 0$. Thus $\alpha^\perp = 0$ characterizes the total geodesic property of $L^\perp$, even in the noninvolutive case (Reinhart [RE 7,8]).

To the bilinear form $\alpha^\perp$ we can further associate a linear map $W^\perp(X) : L^\perp \to L^\perp$ for $X \in \Gamma L$ by the formula

(6.11) $\quad g_L(\alpha^\perp(Z,Z'),X) = g_Q(W^\perp(X)Z,Z')$.

Since $\alpha^\perp$ is symmetric, $W^\perp$ is self-adjoint. Using (6.8) and the torsion-freeness of ∇^M, this formula can be expressed equivalently by

(6.12) $\quad g_Q(W^\perp(X)Z,Z') = g_M(\nabla^M_Z Z', X) - \frac{1}{2} g_M([Z,Z'],X)$
$\qquad\qquad\qquad\qquad = - g_M(Z', \nabla^M_Z X) - \frac{1}{2} g_M([Z,Z'],X),$

but cannot be solved for $W^\perp(X)$, unless $L^\perp$ is involutive. In that case the resulting formula $W^\perp(X)Z = - \pi(\nabla^M_Z X)$ corresponds to (6.4). But even in the noninvolutive case the noninvolutivity terms disappear in the quadratic form associated to (6.12). This is used in the trace calculations below.

MEAN CURVATURE. For a foliation $\mathcal{F}$ on (M, g_M) we define

(6.13) $\quad \kappa(Z) = \text{trace } W(Z) \quad \text{for} \quad Z \in \Gamma L^\perp,$

and set $i(X)\kappa = 0$ for $X \in \Gamma L$, so that $\kappa \in \Omega^1(M)$. This is the mean curvature form of $\mathcal{F}$ (or $L \subset TM$). In terms of a local orthonormal frame $E_i (i = 1,\ldots,p)$ of (L, g_L) we find by (6.3)

(6.14) $\quad \kappa(Z) = \sum_{i=1}^{p} g_Q(\alpha(E_i, E_i), Z).$

Dually we define $\tau \in \Gamma L^\perp$ by

(6.15) $\quad \kappa(Z) = g_Q(\tau, Z),$

and then

(6.16) $\quad \tau = \sum_{i=1}^{p} \alpha(E_i, E_i) = \sum_{i=1}^{p} \pi(\nabla^M_{E_i} E_i).$

This is the usual mean curvature vector field except for a factor $1/p$, which has been suppressed throughout these notes.

Assume L to be oriented. $\mathcal{F}$ is then said to be tangentially oriented. The characteristic form $\chi_{\mathcal{F}}$ of $\mathcal{F}$ on the Riemannian manifold (M, g_M) is defined as follows. It is a p-form on M, which evaluated on a local oriented orthonormal frame E_i $(i = 1,\ldots,p)$ of L gives the value 1 (i.e. is the canonical volume associated to g_L), and for arbitrary $Y_1,\ldots,Y_p \in \Gamma TM$ is given by

$$\chi_{\mathcal{F}}(Y_1,\ldots,Y_p) = \det(g_M(Y_i,E_j)_{ij}).$$

Note that $i(Z)\chi_{\mathcal{F}} = 0$ for $Z \in \Gamma L^\perp$. The following formula is due to Rummler [RU1]

(6.17) $\quad \theta(Z)\chi_{\mathcal{F}}|L = - \kappa(Z)\cdot\chi_{\mathcal{F}}|L \quad$ for $Z \in \Gamma L^\perp$.

Since $i(Z)\chi_{\mathcal{F}} = 0$, this says that the p-form $i(Z)d\chi_{\mathcal{F}} + \kappa(Z)\cdot\chi_{\mathcal{F}}$ evaluates to 0 along L.

PROOF. For $X_1,\ldots,X_p \in \Gamma L$ we have

$$(\theta(Z)\chi_{\mathcal{F}})(X_1,\ldots,X_p) = Z\chi_{\mathcal{F}}(X_1,\ldots,X_p) - \sum_{i=1}^{p} \chi_{\mathcal{F}}(X_1,\ldots,\pi^\perp[Z,X_i],\ldots,X_p).$$

Now we evaluate this on a local orthonormal frame $E_1,\ldots,E_p$ of L. The first term on the RHS vanishes. Further in the term involving $\pi^\perp[Z,E_i]$ this vector field can be replaced by its projection to E_i. Since

$$\pi^\perp[Z,E_i] = \sum_{j=1}^{p} g_M([Z,E_i],E_j)E_j,$$

this implies that

$$(\theta(Z)\chi_{\mathcal{F}})(E_1,\ldots,E_p) = -\sum_{i=1}^{p} g_M([Z,E_i],E_i)\cdot \chi_{\mathcal{F}}(E_1,\ldots,E_p).$$

On the other hand by (6.3)(6.4)

$$\kappa(Z) = \sum_{i=1}^{p} g_L(W(Z)E_i,E_i) = -\sum_{i=1}^{p} g_L(\pi^{\perp}(\nabla^M_{E_i}Z),E_i) = -\sum_{i=1}^{p} g_M(\nabla^M_{E_i}Z,E_i)$$
$$= -\sum_{i=1}^{p} g_M(\nabla^M_Z E_i + [E_i,Z],E_i).$$

Since $2g_M(\nabla^M_Z E_i, E_i) = Zg_M(E_i,E_i) = 0$, this yields the formula

(6.18) $\quad \kappa(Z) = \sum_{i=1}^{p} g_M([Z,E_i],E_i).$

This completes the proof of (6.17). ■

For the particular choice of $Z = \tau$ we have by (6.15)

$$\kappa(\tau) = g_Q(\tau,\tau) = |\tau|^2$$

and by (6.17) we find

(6.19) $\quad (\theta(\tau)\chi_{\mathcal{F}})|L = -|\tau|^2 \cdot \chi_{\mathcal{F}}|L$

τ is the direction of steepest change for $\chi_{\mathcal{F}}$ under the transversal flow φ_t of τ. The vanishing of τ means roughly the invariance of $\chi_{\mathcal{F}}$ under transversal flows (the precise meaning being that $\theta(Z)\chi_{\mathcal{F}}$ evaluates to zero on L).

A foliation with vanishing mean curvature is called harmonic. Every leaf of such a foliation is a minimal submanifold of M (see [KN, II, p. 379] for this concept). The term minimal foliation is already in use for foliations with every leaf dense in M, in accordance with the terminology in topological dynamics. The name harmonic foliations has been proposed in [KT6] for the following reason. View the projection $\pi : TM \to Q$ as a Q-valued 1-form, i.e. $\pi \in \Omega^1(M,Q)$. Then $\mathcal{F}$ is a harmonic 1-form precisely when $\kappa = 0$. Namely $d_\nabla \pi = 0$ always holds for the exterior differential

$$d_\nabla : \Omega^r(M,Q) \to \Omega^{r+1}(M,Q),$$

associated to ∇. For the natural adjoint δ_∇ on such forms one finds then $\delta_\nabla \pi = \tau \in \Gamma Q$ (see [KT6, p. 103] for details). The analogy with the harmonic map theory in the sense of Eells-Sampson [ES] is that a foliation defined by a submersion is harmonic iff the map is harmonic.

For a harmonic foliation $\mathcal{F}$ on (M, g_M) (6.17) implies that $i(Z)d\chi_{\mathcal{F}}$ evaluates to zero on L. Under what condition can one conclude $d\chi_{\mathcal{F}} = 0$? We return to the general situation and prove first the following result.

6.20 THEOREM. <u>Let $\mathcal{F}$ be a tangentially oriented foliation on</u> (M, g_M), <u>and $\chi_{\mathcal{F}}$ the characteristic form of</u> L. <u>Assume $L^\perp$ to be involutive. Then</u>

(6.21) $\quad \theta(Z)\chi_{\mathcal{F}} + \kappa(Z)\cdot\chi_{\mathcal{F}} = 0 \quad$ <u>for</u> $\quad Z \in \Gamma L^\perp$.

PROOF. Let $\alpha = \theta(Z)\chi_{\mathcal{F}} + \kappa(Z)\cdot\chi_{\mathcal{F}} \in \Omega^p(M)$. We know by (6.17) that $\alpha|L = 0$. It suffices to show that $i(Z')\alpha = 0$ for $Z' \in \Gamma L^\perp$. Note that

(6.22) $\quad i(Z')\theta(Z) = \theta(Z)i(Z') - i[Z,Z']$

(see the formulas at the end of Chapter 2). Thus

$$i(Z')\alpha = i(Z')(\theta(Z)\chi_{\mathcal{F}} + \kappa(Z)\cdot\chi_{\mathcal{F}})$$
$$= \theta(Z)i(Z')\chi_{\mathcal{F}} - i[Z,Z']\chi_{\mathcal{F}} + \kappa(Z)\cdot i(Z')\chi_{\mathcal{F}}.$$

For involutive $L^\perp$ we have $[Z,Z'] \in L^\perp$, and $i(Z')\alpha = 0$ follows. ∎

6.23 THEOREM. Let $\mathcal{F}$ be as in Theorem 6.20 with involutive $L^\perp$. Then the following conditions are equivalent:
(i) $\kappa = 0$, i.e. $\mathcal{F}$ is harmonic;
(ii) $\theta(Z)\chi_{\mathcal{F}} = 0$ for $Z \in \Gamma L^\perp$;
(iii) $d\chi_{\mathcal{F}} = 0$.

PROOF. (i)⇔(ii) follows by (6.21). (iii)⇒(ii) follows since $i(Z)\chi_{\mathcal{F}} = 0$ for $Z \in \Gamma L^\perp$. It remains to show (ii)⇒(iii). Thus we assume $i(Z)d\chi_{\mathcal{F}} = 0$ and we need to show that $i(X)d\chi_{\mathcal{F}} = 0$ for $X \in \Gamma L$. We observe first that

(6.24) $(i(X)d\chi_{\mathcal{F}})|L = 0$.

Since $\beta = i(X)d\chi_{\mathcal{F}}$ is a p-form, to show (6.24) it suffices to prove that β evaluates to zero on a local orthonormal frame $E_i (i = 1,\ldots,p)$ of L. But

$$\beta(E_1,\ldots,E_p) = d\chi_{\mathcal{F}}(X;E_1,\ldots,E_p) = 0,$$

since X is a linear combination of the E_i's. This proves (6.24).

To show $\beta = i(X)d\chi_{\mathcal{F}} = 0$ it suffices now to prove that moreover $i(Z)\beta = 0$ for $Z \in \Gamma L^\perp$. But

$$i(Z)i(X)d\chi_{\mathcal{F}} = -i(X)i(Z)d\chi_{\mathcal{F}},$$

and $i(Z)d\chi_{\mathcal{F}} = 0$ by assumption. ∎

We note that this discussion applies in particular to tangentially oriented foliations of codimension one.

It may be worthwhile to note, that in this approach the involutivity of L has never been invoked, and in the last discussion only the involutivity of $L^\perp$. This is useful below, when we switch the roles of L and $L^\perp$.

For the subbundle $L^\perp \subset TM$ we can define in complete analogy a 1-form $\kappa^\perp \in \Omega^1(M)$ by

(6.25) $\quad \kappa^\perp(X) = \text{trace } W^\perp(X) \quad \text{for } X \in \Gamma L,$

and $i(Z)\kappa^\perp = 0$ for $Z \in \Gamma L^\perp$. For the dual vectorfield $\tau^\perp \in \Gamma L$ defined by

(6.26) $\quad \kappa^\perp(X) = g_L(\tau^\perp, X)$

we have then in terms of a local orthonormal frame E_γ ($\gamma = 1,\ldots,q$) of $L^\perp$ the formula

(6.27) $\quad \tau^\perp = \sum_{\gamma=1}^{q} \alpha^\perp(E_\gamma, E_\gamma) = \sum_{\gamma=1}^{q} \pi^\perp(\nabla^M_{E_\gamma} E_\gamma).$

Corresponding to (6.17) we have the following formula. Assume $\mathcal{F}$ to be transversally oriented, and let ν be the characteristic form of $L^\perp \cong \mathbf{Q}$. Then

(6.28) $(\theta(X)\nu)|Q = - \kappa^\perp(X)\cdot\nu|Q$ for $X \in \Gamma L$.

The proof is identical to the proof of (6.17), but carried out in terms of an orthonormal frame E_γ ($\gamma = 1,\ldots,q$) of $L^\perp$. But observe that in this situation L (which now plays the role of the complement of $L^\perp$) is involutive. Thus we can apply (6.21), which proves

(6.29) $\theta(X)\nu + \kappa^\perp(X)\cdot\nu = 0$ for $X \in \Gamma L$.

For the mean curvature $\kappa^\perp$ we obtain the formula corresponding to (6.18)

(6.30) $\kappa^\perp(X) = \sum_{\gamma=1}^{q} \alpha^\perp(\ g_M([X,E_\gamma],E_\gamma)$.

The steepest change of ν_Q is obtained for $x = \tau^\perp$ and then

(6.31) $\theta(\tau^\perp)\nu = -|\tau^\perp|^2\cdot\nu$.

Note that the invariance of ν under tangential flows is precisely the condition (5.6) of holonomy invariance. Corresponding to Theorem 6.23 we have therefore the following result.

6.32 THEOREM. Let $\mathcal{F}$ be a transversally oriented foliation on a Riemannian manifold (M,g_M), and ν the characteristic form of $L^\perp \cong Q$, i.e. the canonical transversal volume form. Then the following conditions are equivalent:
(i) $\kappa^\perp = 0$;
(ii) ν is holonomy invariant, i.e. $\mathcal{F}$ is an SL(q)-foliation;
(iii) $d\nu = 0$.

The proof follows from (6.29). The point is that the arguments presented before for the proof of $d\chi_{\mathcal{F}} = 0$ depend only on the involutivity of $L^{\perp}$, while the involutivity of L was not used. Thus it applies to the situation when L and $L^{\perp}$ are switched.

The conditions in Theorem 6.32 hold in particular for a transversally oriented Riemannian foliation, since the holonomy invariancw of g_Q implies the holonomy invariance of ν. Foliations with holonomy invariant transversal measure have been studied extensively in the broader measure theoretic context by Sacksteder [S], Plante [PL] and others. They play an important role in many contexts.

An interesting idea of Plante [PL] is the following. A closed submanifold T of M of dimension q, at every point transversal to the foliation $\mathcal{F}$ of codimension q with holonomy invariant measure ν, defines a nontrivial homology class $[T] \in H_q(M)$. The reason is that $\nu|T$ defines a closed form with $\int_T \nu > 0$. If $[T] = 0$, then $T = \partial S$ would be the boundary of a $(q+1)$-chain S, and $\int_T \nu = \int_S d\nu = 0$, a contradiction. The case $q = 1$ is discussed in the next chapter.

We finish this chapter by relating the transversal volume form and the characteristic form of a foliation. For this purpose we need the star operator $* : \Omega^r(M) \to \Omega^{n-r}(M)$ associated to the metric g_M. Let M be oriented, and μ the Riemannian volume associated to g_M. Then $*$ is completely characterized by

(6.33) $\quad (*\alpha)(X_1, \ldots, X_{n-r}) \cdot \mu = \alpha \wedge \gamma(X_1) \wedge \cdots \wedge \gamma(X_{n-r})$,

for $\alpha \in \Omega^r(M)$ and vector fields $X_1, \ldots, X_{n-r}$. Here $\gamma : \mathcal{V}(M) \to \Omega^1(M)$ denotes the isomorphism from vector fields to 1-forms defined by the metric.

In particular, $*\mu = 1$ and $*1 = \mu$. For a positively oriented orthonormal frame $E_1,\ldots,E_n$ and its dual coframe $\omega_1,\ldots,\omega_n$ we have

$$(*\alpha)(E_{i_1},\ldots,E_{i_r})\cdot \mu = \alpha \wedge \omega_{i_1} \wedge \cdots \wedge \omega_{i_r}.$$

Let $\mathcal{F}$ be a tangentially and transversally oriented foliation on (M,g_M). For a local orthonormal frame with $E_1,\ldots,E_p \in \Gamma L$ and $E_{p+1},\ldots,E_n \in \Gamma L^\perp$ we find then for the characteristic form ν of $L^\perp$

$$(*\nu)(E_1,\ldots,E_p)\cdot \mu = (\omega_{p+1} \wedge \cdots \wedge \omega_n) \wedge (\omega_1 \wedge \cdots \wedge \omega_p).$$

For $E_\alpha \in \Gamma L^\perp$ we find $i(E_\alpha)(*\nu) = 0$. Thus $*\nu$ is up to sign the characteristic form $\chi_{\mathcal{F}}$ of L. We choose orientations of $L, L^\perp$ and TM, such that $*\nu$ is precisely the characteristic form $\chi_{\mathcal{F}}$ of L. With this convention

(6.34) $\quad \nu \wedge \chi_{\mathcal{F}} = \nu \wedge *\nu = \mu.$

CHAPTER 7
CODIMENSION ONE FOLIATIONS

Let $\mathcal{F}$ be a transversally oriented foliation of codimension one on (M^{n+1}, g_M). Let $Z \in \Gamma L^\perp$ be a unit vector field and $\nu \in \Omega^1(M)$ the dual form, defined by

$$\nu(Y) = g_M(Z,Y) \quad \text{for} \quad Y \in \Gamma TM.$$

ν is then of unit length in the induced (pointwise) metric on 1-forms. $\mathcal{F}$ is defined by ν, so ν is completely integrable. ν is a transversal volume form. The induced metric g_Q and ν are related by

$$g_Q(\lambda Z, \lambda Z) = \lambda^2, \quad \nu(\lambda Z) = \lambda.$$

Let as before ∇^M denote the Levi-Cività connection of M, ∇ the connection in Q defined by g_Q (see (5.3)), and $\overset{\circ}{\nabla}$ its restriction to L. By (6.26) we have

(7.1) $$\kappa^\perp(X) = g_M([X,Z],Z) \quad \text{for} \quad X \in \Gamma L,$$

and by (6.27)

$$\tau^\perp = \pi^\perp(\nabla^M_Z Z).$$

But $g_M(\nabla^M_Z Z, Z) = \frac{1}{2} Z g_M(Z,Z) = 0$, so that $\nabla^M_Z Z \in \Gamma L$. It follows that

(7.2) $$\tau^\perp = \nabla^M_Z Z \in \Gamma L.$$

With those notations we have then the following characterizations of Riemannian foliations of codimension $q = 1$ [KT 6,11].

7.3 THEOREM. If $q = 1$ the following conditions for a transversally oriented foliation on (M, g_M) are equivalent:

(i) $\mathcal{F}$ is Riemannian, and g_M a bundle-like metric;

(ii) $d\nu = 0$;

(iii) $\theta(X)\nu \equiv \overset{\circ}{\nabla}{}^*_X \nu = 0$ for $X \in \Gamma L$;

(iv) $\overset{\circ}{\nabla}_X Z = 0$ for $X \in \Gamma L$;

(v) $\tau^\perp \equiv \nabla^M_Z Z = 0$;

(vi) $\nabla Z = 0$

(vii) $\nabla^M_Z X \in \Gamma L$ for $X \in \Gamma L$;

(viii) $\theta(Z)\nu = 0$.

PROOF. By Corollary 6.10, condition (i) is characterized by $\alpha^\perp = 0$. By Theorem 6.32

$$\kappa^\perp = 0 \Leftrightarrow \theta(X)\nu = 0 \Leftrightarrow d\nu = 0.$$

By (5.4) we have $\theta(X)\nu = \overset{\circ}{\nabla}{}^*_X \nu$, and further $(\overset{\circ}{\nabla}{}^*_X \nu)(Z) = X\nu(Z) - \nu(\overset{\circ}{\nabla}_X Z) = -\nu(\overset{\circ}{\nabla}_X)$. For $q = 1$ therefore $\kappa^\perp = 0 \Leftrightarrow \tau^\perp = 0 \Leftrightarrow \alpha^\perp = 0$. These observations show (i) $\Leftrightarrow$ (ii) $\Leftrightarrow$ (iii) $\Leftrightarrow$ (iv) $\Leftrightarrow$ (v). The equivalence of (vi) with (iv) and (v) follows from the definition (5.3) of ∇. The equivalence (v) $\Leftrightarrow$ (vii) follows from $\nabla^M_Z Z \in \Gamma L$ and

$$g_M(\nabla^M_Z Z, X) = Z g_M(Z, X) - g_M(Z, \nabla^M_Z X) = -g_M(Z, \nabla^M_Z X).$$

It remains to show the equivalence of these conditions with (viii).

Let $\alpha = -\theta(Z)\nu$. Then by Proposition 2.3 we have $d\nu = \alpha \wedge \nu$, which shows (viii) ⇒ (ii). Assume conversely $d\nu = 0$. Then

$$0 = d\nu(X,Z) = (\alpha \wedge \nu)(X,Z) = \alpha(X) \cdot \nu(Z) - \alpha(Z) \cdot \nu(X) = \alpha(X).$$

Thus $\alpha|L = 0$. Since $\alpha|L^\perp = 0$, this shows that $\alpha = 0$. ∎

The form $\alpha = -\theta(Z)\nu$ leads by Proposition 2.3 to the Godbillon-Vey class $[\alpha \wedge d\alpha]$ of $\mathcal{F}$. The identity $d\nu = \alpha \wedge \nu$ reads

(7.4) $$d\nu + \theta(Z)\nu \wedge \nu = 0.$$

We wish to compare this with the identity (6.29), i.e.

(7.5) $$\theta(X)\nu + \kappa^\perp(X) \cdot \nu = 0 \quad \text{for } X \in \Gamma L.$$

The identities (7.4) and (7.5) are linked by

(7.6) $$\kappa^\perp = \theta(Z)\nu,$$

i.e. $\alpha = -\kappa^\perp$ is a form realizing the Godbillon-Vey class $[\alpha \wedge d\alpha]$ of $\mathcal{F}$.

PROOF of (7.6). Since $\kappa^\perp(Z) = 0$ by definition, and $(\theta(Z)\nu)(Z) = Z\nu(Z) - \nu(\theta(Z)Z) = 0$, it suffices to verify

(7.7) $$\kappa^\perp(X) = i(X)\theta(Z)\nu.$$

By (6.30) we have $\kappa^\perp(X) = g_M([X,Z], Z)$. On the other hand

$$i(X)\theta(Z)\nu = \theta(Z)i(X)\nu - i[Z,X]\nu.$$

But $i(X)\nu = 0$ and, since ν is the 1-form dual to Z,

$$\nu([Z,X]) = g_M([Z,X],Z),$$

which proves (7.7). ∎

The operator $i(X)$ applied to (7.4) implies (7.5). But conversely (7.5) also implies (7.4). Namely using (7.6), the identity (7.5) takes the form

$$\theta(X)\nu + i(X)\theta(Z)\nu \cdot \nu = 0,$$

or equivalently

$$i(X)(d\nu + \theta(Z)\nu \wedge \nu) = 0.$$

To prove (7.4), it suffices to verify that $i(Z)\gamma = 0$ for $\gamma = d\nu + \theta(Z)\nu \wedge \nu$. But

$$i(Z)\gamma = i(Z)d\nu + i(Z)\theta(Z)\nu \cdot \nu - \theta(Z)\nu \cdot i(Z)\nu$$

$$= \theta(Z)\nu + \theta(Z)i(Z)\nu \cdot \nu - \theta(Z)\nu = 0.$$

Thus (7.4) and (7.5) are equivalent.

Examples of such foliations are given by invariant 1-forms ν on a symmetric space. They are necessarily closed, and nonsingular if nontrivial. In fact the pointwise norm $|\nu|$ with respect to an invariant metric is constant. Thus the space of Riemannian foliations so obtained is $\Omega^1(M)^G/\mathbb{R}$, if $M = G/H$, where $\Omega^1(M)^G$ denotes the space of G-invariant 1-forms on M. For the torus T^{n+1} this leads to the linear foliations given by 1-forms with constant coefficients (and pointwise norm 1).

If on the other hand $\pi_1 M = 0$, or more generally $H^1_{DR}(M) = 0$, for a closed manifold M, foliations of codimension one on M are not Riemannian. Namely let $\mathcal{F}$ be such a foliation. A covering argument shows that we can assume $\mathcal{F}$ to be transversally oriented. By Theorem 7.3 there is then a closed form ν defining $\mathcal{F}$. Since $H^1_{DR}(M) = 0$ by assumption, $\nu = df$ for some $f : M \longrightarrow \mathbb{R}$. The critical points of f are singularities of ν and $\mathcal{F}$, a contradiction.

Note that for closed M we have $H^1(M,\mathbb{R}) \cong \text{Hom}(H_1(M),\mathbb{R})$. To see that the existence of a transversally oriented Riemannian foliation $\mathcal{F}$ on M implies $H_1(M) \neq 0$, we can alternatively argue as follows. Let $\gamma : [a,b] \longrightarrow M$ be a curve such that $\dot{\gamma}(t) \in L^\perp_{\gamma(t)}$ for all $t \in [a,b]$. Such a curve is called horizontal. Then for the 1-form ω defining $\mathcal{F}$ as in Theorem 7.3 we have

$$\int_\gamma \nu = \int_a^b \nu(\dot{\gamma}(t))dt = \int_a^b |\dot{\gamma}(t)|dt = \ell(\gamma) > 0,$$

where $\ell(\gamma)$ is the length of γ. Assume now that $\gamma : S^1 \longrightarrow M$ is closed. Then $\partial \gamma = 0$ and γ represents a homology class $[\gamma] \in H_1(M)$. We claim that $[\gamma] \neq 0$. For if $[\gamma] = 0$, then the curve γ bounds a 2-dimensional smooth chain c, $\gamma = \partial c$, and

$$\ell(\gamma) = \int_\gamma \nu = \int_{\partial c} \nu = \int_c d\nu = 0,$$

since $\mathcal{F}$ is Riemannian. This contradicts $\ell(\gamma) > 0$.

To make this argument conclusive, it suffices to prove the existence of a closed curve γ, which at every point $\gamma(t)$ is transversal to a given transversally oriented foliation $\mathcal{F}$ of codimension one on a closed manifold M^{n+1}. This suffices to establish $\int_\gamma \nu \neq 0$, even if $\nu(\dot{\gamma}(t))$ does not necessarily coincide with $|\dot{\gamma}(t)|$.

We consider the orbit γ of a point $x \in M^{n+1}$ under the flow φ_t of Z. If γ is not yet closed, we can modify it to a closed γ' as in the preceding paragraph by the following argument. By compactness there is a distinguished chart $(x,y) : U \longrightarrow \mathbb{R}^n \times \mathbb{R}$ such that γ passes through the disk D^n with $|x| \leq 1$ and $y = 0$ infinitely often. A segment of γ which begins and ends on D^n with close endpoints, can then be modified so that the resulting γ' is closed, and still transversal to $\mathcal{F}$ at every point $\gamma'(t)$.

We note that the 1-cycle so constructed yields for a Riemannian $\mathcal{F}$ a nontrivial integral homology class of M^{n+1}.

Next we wish to examine the characteristic form of a codimension one foliation. We assume $\mathcal{F}$ to be tangentially and transversally oriented. With the compatible orientation on M explained at the end of Chapter 6 we have then for the star operator associated to g_M

$$(7.8) \qquad \chi_{\mathcal{F}} = {}^*\nu \quad \text{and} \quad \nu \wedge \chi_{\mathcal{F}} = \mu,$$

where $\chi_{\mathcal{F}}$ is the characteristic form of $\mathcal{F}$. This implies in view of $i(Z)\nu = 1$ and $i(Z)\chi_{\mathcal{F}} = 0$ the formula

$$(7.9) \qquad \chi_{\mathcal{F}} = i(Z)\mu.$$

We further wish to show that the formula (6.21) is in this case equivalent to the identity

(7.10) $$d\chi_{\mathcal{F}} + \kappa \wedge \chi_{\mathcal{F}} = 0.$$

Thus formally the form κ plays, up to sign, for the one dimensional foliation $\mathcal{F}^{\perp}$ the role of the form α leading to the Godbillon-Vey class $\alpha \wedge (d\alpha)^n$. But note that the form $\kappa \wedge (d\kappa)^n$ is a $(2n+1)$-form, thus invisible on M^{n+1}.

PROOF of (7.10). $i(Z)$ applied to (7.10) yields (6.21). To prove that (6.21) implies (7.10), we write (6.21) in the form

$$i(Z)(d\chi_{\mathcal{F}} + \kappa \wedge \chi_{\mathcal{F}}) = 0,$$

which follows from $i(Z)\chi_{\mathcal{F}} = 0$. It suffices to show that the p-form

$$\beta = i(X)(d\chi_{\mathcal{F}} + \kappa \wedge \chi_{\mathcal{F}})$$

vanishes. For this it suffices to show that $\beta|L = 0$ and $i(Z)\beta = 0$. The last fact is clear, since

$$i(Z)\beta = i(Z)i(X)(d\chi_{\mathcal{F}} + \kappa \wedge \chi_{\mathcal{F}}) = -i(X)i(Z)(d\chi_{\mathcal{F}} + \kappa \wedge \chi_{\mathcal{F}}) = 0.$$

To prove $\beta|L = 0$, we evaluate it on a (local) orthonormal frame $E_1, \ldots, E_n$ ($p = n$) of L. Clearly $d\chi_{\mathcal{F}}(X; E_1, \ldots, E_n) = 0$, since X is a linear combination of the E_i's. Further $(\kappa \wedge \chi_{\mathcal{F}})(X; E_1, \ldots, E_n) = 0$, since $\kappa|L = 0$ by definition. ∎

From (7.9) we have

$$d\chi_{\mathcal{F}} = di(Z)\mu = \theta(Z)\mu = \text{div}Z.\mu = \text{div } Z \cdot (\nu \wedge \chi_{\mathcal{F}})$$

so that (7.10) can also be written in the form

$$(\text{div } Z \cdot \nu + \kappa) \wedge \chi_{\mathcal{F}} = 0.$$

Since $i(X)\nu = 0$, $i(X)\kappa = 0$, and $i(Z)\chi_{\mathcal{F}} = 0$, this implies the formula

(7.11) $$\kappa = - \text{div } Z \cdot \nu.$$

7.12 EXAMPLE. We consider the foliation of $\mathbb{R}^{n+1} - \{0\}$ by concentric spheres. The leaves are given as the level hypersurfaces of the function $f(x) = \frac{1}{2}|x|^2 = \frac{1}{2}\sum_{i=1}^{n+1} x_i^2$ for $x = (x_1, \ldots, x_{n+1})$. Since the leaves are the nontrivial orbits of $SO(n+1)$ acting on $\mathbb{R}^{n+1}$, this foliation is Riemannian. Let N be the radial vector field given by

$$N_x = x = \sum_{i=1}^{n+1} x_i \frac{\partial}{\partial x_i} = (\nabla f)_x$$

for $x = (x_1, \ldots, x_{n+1})$. Since $|N_x| = |x| = r$, it follows that

(7.13) $$Z = \frac{1}{r} N$$

is the unit normal vector field to $\mathcal{F}$. The dual 1-form ν is given by

$$\nu(X) = \langle Z, X \rangle = \frac{1}{r}\langle N, X \rangle = \frac{1}{r} df(X).$$

Thus $\nu = \frac{1}{r} df$, where $df = \sum_{i=1}^{n+1} x_i dx_i$. Note that $f = \frac{1}{2} r^2$, so that $df = rdr$ and $\nu = dr$. This shows $d\nu = 0$, which by part (ii) of Theorem 7.3 proves again that $\mathcal{F}$ is Riemannian.

By part (iv) of Theorem 7.3 this is equivalent to $\overset{\circ}{\nabla}_X Z = 0$ for $X \in \Gamma L$. It follows that

$$(7.14) \qquad \overset{\circ}{\nabla}_X N = \overset{\circ}{\nabla}_X (rZ) = Xr \cdot Z = 0,$$

since $X \in \Gamma L$ means that X is tangent to the spheres r = constant.

Let $\widetilde{\nabla}$ denote the canonical connection of $\mathbb{R}^{n+1}$, the Levi-Cività connection of the standard metric on $\mathbb{R}^{n+1}$. Then clearly $\widetilde{\nabla}_Z Z = 0$ (see also part (v) of Theorem 7.3). For the induced connection ∇ in the normal line bundle, we claim that

$$(7.15) \qquad \nabla_N N = \widetilde{\nabla}_N N = N.$$

PROOF of (7.15). For $N = rZ$ we have

$$\widetilde{\nabla}_N N = \widetilde{\nabla}_{rZ} N = r(\widetilde{\nabla}_Z (rZ)) = r(Zr \cdot Z),$$

since $\widetilde{\nabla}_Z Z = 0$. For $Z = \frac{1}{r} \sum x_i \frac{\partial}{\partial x_i}$ we find, in view of $\frac{\partial r}{\partial x_i} = \frac{x_i}{r}$, indeed $Zr = 1$, which shows $\widetilde{\nabla}_N N = N$. Since the resulting vector is already normal, the same holds for $\nabla_N N = \pi(\widetilde{\nabla}_N N)$. ∎

Next we determine the Weingarten map $W(N) : L \longrightarrow L$, and show

$$(7.16) \qquad W(N) = -\text{id}.$$

PROOF of (7.16). Recall that $W(N)X = -\pi^\perp(\tilde{\nabla}_X N)$, so that it suffices to show $\pi^\perp(\tilde{\nabla}_X N) = X$. Since

$$\langle \tilde{\nabla}_X N, N \rangle = \frac{1}{2} X \langle N, N \rangle = 0,$$

it follows that $\tilde{\nabla}_X N \in \Gamma L$. Thus (7.16) is equivalent to

(7.17) $\qquad\qquad \tilde{\nabla}_X N = X \quad \text{for} \quad X \in \Gamma L.$

PROOF of (7.17). Since $\tilde{\nabla}_X N$ is linear in X, it suffices to show this for a set of vector fields spanning L_x at every $x \in \mathbb{R}^{n+1} - \{0\}$. Since $L_x = T_x S(r)$ ($S(r)$ = sphere of radius $r = |x|$), this is the case for the vector fields corresponding to 1-parameter subgroups of $SO(n+1)$. They are the linear vector fields

$$X^A : x \longmapsto Ax$$

corresponding to skew-symmetric linear maps $A : \mathbb{R}^{n+1} \longrightarrow \mathbb{R}^{n+1}$. Let $A = (a_{ij})$. Then

$$X^A_x = \sum_{ij} a_{ij} x_i \frac{\partial}{\partial x_j}$$

so that applied to the coordinate function x_k we obtain $X^A(x_k) = \sum_{ij} a_{ij} x_i \delta_{jk} = \sum_i a_{ik} x_i$. It follows that for $X = X^A$

$$\tilde{\nabla}_X N = \tilde{\nabla}_X \left[\sum_k x_k \frac{\partial}{\partial x_k} \right] = \sum_k X(x_k) \cdot \frac{\partial}{\partial x_k},$$

since $\tilde{\nabla}_X \frac{\partial}{\partial x} = 0$. Thus

$$\nabla_X N = \sum_{ik} a_{ik} x_i \frac{\partial}{\partial x_k} = X.$$

This proves (7.17). ∎

As a consequence of (7.16) we find for the mean curvature form

(7.18) $\qquad \kappa(N) = \text{trace } W(N) = -n,$

and $\kappa(Z) = -\frac{n}{r}$. For the mean curvature vector field τ, the identity $\kappa(Z) = \langle \tau, Z \rangle$ shows therefore

(7.19) $\qquad \tau = -\frac{n}{r} Z = -\frac{n}{r^2} N.$

The conventional normalization would be $\frac{1}{n}\tau$, but we have suppressed the factor $\frac{1}{n}$. This formula implies

(7.20) $\qquad \kappa = -\frac{n}{r} \cdot \nu.$

To check that this formula corresponds to (7.11), we calculate $\text{div } Z = \sum_i \frac{\partial a_i}{\partial x_i}$ for $Z = \sum_i a_i \frac{\partial}{\partial x_i} = \frac{1}{r} \sum_i x_i \frac{\partial}{\partial x_i}$. But

$$\frac{\partial}{\partial x_i}\left[\frac{x_i}{r}\right] = \frac{1}{r^2}\left[r - x_i \cdot \frac{\partial r}{\partial x_i}\right] = \frac{1}{r^2}\left[r - \frac{1}{r} \cdot x_i^2\right],$$

so that indeed

(7.21) $\qquad \text{div } Z = \frac{n}{r} \quad \text{and} \quad \kappa = -\text{div } Z \cdot \nu.$

Next we note that $\tau \in \Gamma L^\perp$ is a parallel section along L, i.e.

(7.22) $$\overset{\circ}{\nabla}_X \tau = 0 \quad \text{for} \quad X \in \Gamma L.$$

PROOF of (7.22). We have

$$\overset{\circ}{\nabla}_X \tau = X\left[-\frac{n}{r^2}\right] \cdot N + \left[-\frac{n}{r^2}\right] \cdot \overset{\circ}{\nabla}_X N.$$

Using (7.14), it suffices to verify $X\left[-\frac{n}{r^2}\right] = 0$ for $X \in \Gamma L$. But X_x is tangent to the sphere $S(r)$, $r = |x|$, so this is clear. ∎

This does not mean that τ is ∇-parallel along normal directions. In fact

(7.23) $$\nabla_N \tau = \overset{\sim}{\nabla}_N \tau = \frac{n}{r^2} \cdot N \equiv \frac{n}{r} \cdot Z.$$

PROOF of (7.23). By (7.19) and (7.15) we have

$$\overset{\sim}{\nabla}_N \tau = \overset{\sim}{\nabla}_N \left[-\frac{n}{r^2} \cdot N\right] = N\left[-\frac{n}{r^2}\right] \cdot N - \frac{n}{r^2} \cdot N.$$

But $N\left[\frac{1}{r^2}\right] = \sum_i x_i \frac{\partial}{\partial x_i}\left[\frac{1}{r^2}\right] = -\frac{2}{r^2}$, so that

$$\overset{\sim}{\nabla}_N \tau = \left[\frac{2n}{r^2} - \frac{n}{r^2}\right] \cdot N = \frac{n}{r^2} \cdot N.$$

Since the result is already normal, the same holds for $\nabla_N \tau = \pi(\overset{\sim}{\nabla}_N \tau)$. ∎

To determine the characteristic form of $\mathcal{F}$, we consider the volume $\mu = dx_1 \wedge \cdots \wedge dx_{n+1}$. By (7.9), we find then for $Z = \frac{1}{r} \sum_{j=1}^{n+1} x_j \frac{\partial}{\partial x_j}$

(7.24) $\quad \chi_{\mathcal{F}} = i(Z)\mu = \frac{1}{r} \sum_{j=1}^{n+1} (-1)^{j-1} x_j dx_1 \wedge \cdots \wedge \hat{dx_j} \wedge \cdots \wedge dx_{n+1}.$

Since the star operator $* : \Omega^1(\mathbb{R}^{n+1}) \longrightarrow \Omega^n(\mathbb{R}^{n+1})$ is given by

$$*\left[\sum_i a_i dx_i\right] = \sum_i (-1)^{i-1} a_i dx_1 \wedge \cdots \wedge \hat{dx_i} \wedge \cdots \wedge dx_{n+1},$$

it follows that indeed $*\nu = \chi_{\mathcal{F}}$ for

$$\nu = dr = \frac{1}{r} df = \frac{1}{r} \sum_{i=1}^{n+1} x_i dx_i.$$

The form $\chi_{\mathcal{F}}$ restricts on each sphere $S^n(r)$ to its standard volume form, i.e.

$$\text{Vol } S^n(r) = \int_{S^n(r)} i(Z)\mu.$$

From (7.24) we find that

(7.25) $\quad\quad\quad\quad d(r \cdot \chi_{\mathcal{F}}) = (n + 1)\mu,$

so that the volume of the ball $B^{n+1}(r)$ is given by

$$\text{Vol } B^{n+1}(r) = \int_{B^{n+1}(r)} \mu = \frac{1}{n+1} \int_{B^{n+1}(r)} d(r\chi_{\mathcal{F}}) = \frac{r}{n+1} \int_{S^n(r)} \chi_{\mathcal{F}}.$$

This yields the classical formula

$$\text{Vol } B^{n+1}(r) = \frac{r}{n+1} \text{ Vol } S^n.$$

It is worth noting that

(7.26) $$d\left[\frac{1}{r^n} \cdot \chi_{\mathcal{F}}\right] = 0 \quad \text{in } \mathbb{R}^{n+1} - \{0\}.$$

PROOF of (7.26). We calculate more generally

$$d\left[\frac{1}{r^k} \cdot r\chi_{\mathcal{F}}\right] = -k \cdot \frac{1}{r^{k+1}} dr \wedge r\chi_{\mathcal{F}} + \frac{1}{r^k} d\left[r\chi_{\mathcal{F}}\right].$$

Using $dr = \frac{1}{r} \sum_i x_i dx_i$ and (7.24) we find

$$dr \wedge r\chi_{\mathcal{F}} = \frac{1}{r} \cdot \sum_i x_i^2 \cdot \mu = r\mu,$$

and thus by (7.25)

$$d\left[\frac{1}{r^k} \cdot r\chi_{\mathcal{F}}\right] = \frac{1}{r^k}\left[-k + (n+1)\right] \cdot \mu$$

which vanishes (precisely) for $k = n + 1$. ∎

Since

$$\int_{S^n(r)} \frac{1}{r^n} \cdot \chi_{\mathcal{F}} = \frac{1}{r^n} \cdot \text{Vol } S^n(r) \neq 0 \;,$$

it follows that the De Rham class

(7.27) $$\left[\frac{1}{r^n} \cdot \chi_{\mathcal{F}}\right] \neq 0 \in H^n(\mathbb{R}^{n+1} - \{0\}).$$

In $\mathbb{R}^3 - \{0\}$, we have the nontrivial De Rham class represented by

$$\frac{1}{r^2} \chi_{\mathcal{F}} = \frac{1}{r^3} (x_1 \cdot dx_2 \wedge dx_3 - x_2 \cdot dx_1 \wedge dx_3 + x_3 \cdot dx_1 \wedge dx_2),$$

and which on $S^2(r)$ integrates to 4π. Thus

(7.28) $$\left[\frac{1}{2\pi r^2} \cdot \chi_{\mathcal{F}}\right] \in H^2(\mathbb{R}^3 - \{0\}, \mathbb{Z})$$

is an integral cohomology class, evaluating on each $S^2(r)$ to the Euler number $\chi(S^2) = 2$. The integrality implies that $\frac{1}{r^2} \cdot \chi_{\mathcal{F}}$ is the curvature of a connection in a 2-plane bundle, representing 2π times the Euler class of this 2-plane bundle. This is precisely the tangent bundle L of $\mathcal{F}$, restricting on each leaf $S(r)$ to the tangent bundle $TS(r)$.

7.29 REMARK. For a (fictional) magnetic monopole of strength g located at the origin, the magnetic force field in $\mathbb{R}^3 - \{0\}$ is given by

$$B_x = \frac{g}{r^3} \cdot x = \frac{g}{r^3} \sum_{i=1}^{3} x_i \frac{\partial}{\partial x_i} \;.$$

The dual 1-form is then (with the notation at the bottom of p. 72)

$$\gamma(B)_x = \frac{g}{r^3} \sum_i x_i dx_i.$$

The corresponding Maxwell field $F = *\gamma(B)$ is therefore

$$F = \frac{g}{r^2} \cdot \chi_{\mathcal{F}}.$$

As a consequence of (7.26) we have

$$dF = 0,$$

which is one of Maxwell's equations for the magnetostatic field generated by the magnetic monopole.

The cohomology class of F is nontrivial, since

$$\left[\frac{1}{2\pi g} \cdot F\right] \in H^2(\mathbb{R}^3 - \{0\}, \mathbb{Z})$$

is the integral cohomology class discussed before. This is the reason for the nonexistence of a (global) magnetic vector potential in $\mathbb{R}^3 - \{0\}$, in a naive sense. It does exist as a connection in a plane bundle (determined by the integral cohomology class above), and F is the curvature of this connection.

These remarks are a crucial point in Dirac's discussion of quantization conditions in [DI]. Dirac argues that an electron e moving in the magnetic field of the monopole g necessarily satisfies the integrality condition $2eg \in \mathbb{Z}$. This is a consequence of the geometric-topological fact above, and the Schrödinger equation for the wave function of the electron.

Returning to the geometric context, and using the Euler class of the bundle L, we prove the following result of Ehresmann and Reeb [ER].

7.30 THEOREM. Let M^3 be an oriented closed 3-manifold with finite fundamental group. Then a closed leaf of a transversally oriented codimension one foliation on M^3 is necessarily a torus T^2.

PROOF. Let $e(L)$ be the Euler class of the (oriented) tangent bundle L of the foliation, $e(L) \in H^2(M)$. The assumption on $\pi_1 M$ implies $H_1 M = 0$. By Poincaré duality it follows that $H^2 M = 0$. Let $\chi \in \Omega^2(M)$ be a closed 2-form representing the Euler class. It follows that $\chi = d\alpha$, $\alpha \in \Omega^1(M)$. For a closed (oriented) leaf $\mathcal{L}$ of $\mathcal{F}$ we have then

$$\int_{\mathcal{L}} \chi = \int_{\mathcal{L}} d\alpha = \int_{\partial \mathcal{L}} \alpha = 0.$$

But the LHS is the Euler characteristic of $\mathcal{L}$, hence that $\mathcal{L}$ is a 2-torus. ∎

We further prove a related result in arbitrary codimension [ER].

7.31 THEOREM. Let $\mathcal{F}$ be a transversally oriented foliation of an open contractible subset of $\mathbb{R}^{n+1}$ or S^{n+1} (n even). Then a closed leaf of $\mathcal{F}$ has Euler characteristic zero.

PROOF. Let $\mathcal{L}$ be a closed leaf. With no loss of generality, we can assume $\mathcal{L}$ to be oriented. In case the dimension p of the leaves is odd there is nothing to prove. Thus let p be even. Let N be the normal bundle of $\mathcal{L}$, and SN its bundle of unit vectors. The Gauss map $f : SN \longrightarrow S^n$ is then well defined, and its degree according to Hopf given by

$$\deg f = \frac{1}{2} \chi(SN).$$

On the other hand $SN \longrightarrow \mathcal{L}$ is a sphere bundle with fibers of dimension n - p, which is even. Thus

$$\chi(SN) = \chi(S^{n-p}) \cdot \chi(\mathcal{L}) = 2\chi(\mathcal{L}).$$

It follows that $\chi(\mathcal{L})$ = deg f. But radially shrinking SN gives a homotopy of f to a map, which does not cover S^n. Thus deg f = 0 and $\chi(\mathcal{L}) = 0$. ∎

Next we consider the codifferential $\delta \nu = - *d*\nu$ of the transversal volume ν. In a (local) orthonormal frame $E_1,\ldots,E_{n+1}$ of TM^{n+1}

$$(7.32) \qquad \delta \nu = - \sum_{A=1}^{n+1} (\nabla_{E_A} \nu)(E_A) = - \sum_{A=1}^{n+1} \left[E_A \nu(E_A) - \nu(\nabla_{E_A}^M E_A) \right].$$

Assume $E_{n+1} = Z$. Then $E_1,\ldots,E_n$ is a local orthonormal frame of L. Since $\nu(E_i) = 0$ for $i = 1,\ldots,n$ and $\nu(Z) = 1$, it follows that the first terms on the RHS vanish. Moreover by (7.2) the vector field $\nabla_Z^M Z$ is tangential to $\mathcal{F}$, so that $\nu(\nabla_Z^M Z) = 0$. It follows that

$$(7.33) \qquad \delta \nu = \sum_{i=1}^{n} \nu(\nabla_{E_i}^M E_i).$$

But $\nu(\nabla_{E_i}^M E_i) = g_M(Z, \nabla_{E_i}^M E_i) = E_i g_M(Z, E_i) - g_M(\nabla_{E_i}^M Z, E_i)$. The first term vanishes, so that

$$\delta \nu = - \sum_{i=1}^{n} g_M(\nabla_{E_i}^M Z, E_i).$$

Comparing this with the calculations leading to (6.18), we find

(7.34) $$\delta\nu = \kappa(Z).$$

We have then the following characterizations of transversally oriented harmonic foliations of codimension $q = 1$ [KT6].

7.35 THEOREM. If $q = 1$, the following conditions for a transversally oriented foliation on (M, g_M) are equivalent:
 (i) $\mathcal{F}$ is harmonic (all leaves are minimal submanifolds);
 (ii) $\delta\nu = 0$;
 (iii) $d\chi_{\mathcal{F}} = 0$ (Rummler's condition);
 (iv) div $Z = 0$.

PROOF. (i) is characterized by $\kappa = 0$. (i) $\Leftrightarrow$ (ii) follows from (7.34). (i) $\Leftrightarrow$ (iii) follows from (7.10). (ii) $\Leftrightarrow$ (iv) finally follows from

(7.36) $$\delta\nu = -\text{div } Z$$

which holds for any vector field Z and its dual 1-form ν. ∎

As an illustration we prove the following result.

7.37 THEOREM. Let M^3 be a closed oriented 3-manifold with finite fundamental group. Then a transversally oriented foliation of codimension one on M is not harmonic.

PROOF. Let Z be a unit normal vector field, ν its dual 1-form and $\chi_{\mathcal{F}} = *\nu$ the characteristic form. If we assume $\mathcal{F}$ harmonic, then $d\chi_{\mathcal{F}} = 0$. By Novikov [N], under the given hypothesis $\mathcal{F}$ has a closed leaf $\mathcal{L}$. Thus $\int_{\mathcal{L}} \chi_{\mathcal{F}} = \text{Vol}(\mathcal{L}) > 0$.

On the other hand we have by Poincaré duality $H^2_{DR}(M) = 0$, so that $\chi_{\mathcal{F}} = d\gamma$ for some $\gamma \in \Omega^1(M)$. This yields

$$0 < \int_{\mathcal{L}} \chi_{\mathcal{F}} = \int_{\mathcal{L}} d\gamma = \int_{\partial \mathcal{L}} \gamma = 0,$$

a contradiction derived from the harmonicity of $\mathcal{F}$. ∎

This argument shows that in fact $\mathcal{F}$ is not harmonic for any metric whatsoever on M. A foliation is called <u>taut</u>, if there is at least one metric for which $\mathcal{F}$ is harmonic. Criteria for tautness have been discussed by Rummler [RU 1,2] and Sullivan [SU 1,2]. The statement of Theorem 7.37 is then that a transversally oriented foliation of codimension one on a closed oriented 3-manifold with finite fundamental group is not taut.

This result applies to the Reeb foliation on S^3, which therefore is not taut. Note that there the closed leaf T^2 appears explicitly as the boundary ∂c of the solid 3-torus c in S^3.

More generally for a (transversally and tangentially orientable) foliation $\mathcal{F}$ on M^n we have $d\chi_{\mathcal{F}} = d*\nu$, so that

(7.38) $\qquad d\chi_{\mathcal{F}} = 0 \Leftrightarrow \delta\nu = (-1)^{n(q+1)+1} * d * \nu = 0.$

The argument above is based on the vanishing of $[\chi_{\mathcal{F}}] \in H^p_{DR}(M)$.

Suppose that $\chi_{\mathcal{F}}$ is closed. Then no closed leaf $\mathcal{L}$ is a boundary of a smooth (p + 1)-chain c. For if $\mathcal{L} = \partial c$, then

(7.39)
$$0 < \int_{\mathcal{L}} \chi_{\mathcal{F}} = \int_{\partial c} \chi_{\mathcal{F}} = \int_{c} d\chi_{\mathcal{F}} = 0,$$

a contradiction. A closed leaf $\mathcal{L}$ thus gives rise to a nontrivial homology class in $H_p M$.

This is dual to Plante's statement mentioned in Chapter 6, namely that for a foliation with holonomy invariant transverse measure ν (and hence $d\nu = 0$), a transversal q-cycle T leads to a nontrivial homology class $[T] \in H_q(M)$. Namely $T = \partial c$ would yield the contradiction

(7.40)
$$0 < \int_T \nu = \int_{\partial c} \nu = \int_c d\nu = 0.$$

Sullivan had the idea to reverse these arguments [SU 2] [SU 3]. Namely the nonexistence of tangential boundary relations for "generalized closed leaves" of a foliation $\mathcal{F}$ can conversely be used to prove the existence of a closed characteristic form $\chi_{\mathcal{F}}$ for $\mathcal{F}$. The basic idea is to view $\chi_{\mathcal{F}}$ as a functional on the space of p-currents of M. The properties of $\chi_{\mathcal{F}}$ are then its vanishing on the closure of the subspace of tangential boundary currents, and the positivity on the separated cone of currents generated by the p-vectors tangent to $\mathcal{F}$. An application of the Hahn-Banach theorem, and of the reflexive duality between currents and forms (Schwartz), leads then conversely to the existence of $\chi_{\mathcal{F}}$. This form can then be realized as the characteristic form arising from a Riemannian metric. This leads to a purely topological characterization of taut foliations by the nonexistence of so called "tangential homologies" [SU 2] [SU 3]. These matters are also discussed in [KT 11].

The characterizations given in this chapter for Riemannian and harmonic foliations of codimension one are encapsulated in the formulas

(7.41) $$d\nu = -\kappa^\perp \wedge \nu$$

(7.42) $$\delta\nu = \kappa(Z).$$

The first of these formulas is (7.4) together with (7.6). The second formula is (7.34). They hold for a transversally oriented foliation with unit normal vectorfield Z, and dual transversal volume 1-form ν.

The following result can be deduced from [SU3].

7.43 THEOREM. Let $\mathcal{F}$ be a transversally oriented foliation of codimension one on a closed manifold M. If $\mathcal{F}$ is Riemannian, then $\mathcal{F}$ is taut.

This is proved in [KT 11, Thm 3.26] by an argument originally due to Calabi [CL]. The point is to modify the metric, so as to make the nonsingular closed 1-form ν coclosed in the new metric. As stated before, such foliations cannot exist if M is simply connected.

Next we analyze $\kappa^\perp$ in more detail. The dual vector field $\tau^\perp$ is then $\tau^\perp = \nabla^M_Z Z$. We show that for $X, X' \in \Gamma L$

(7.44) $$d\kappa^\perp(X,X') = g_M(X', \nabla^M_X \tau^\perp) - g_M(X, \nabla^M_{X'} \tau^\perp).$$

PROOF of (7.44). We have for $\kappa^\perp = \theta(Z)\nu$ (see (7.6))

$$d\kappa^\perp(X,X') = X\kappa^\perp(X') - X'\kappa^\perp(X) - \kappa^\perp[X,X']$$

$$= Xg_M(X', \tau^\perp) - X'g_M(X, \tau^\perp) - g_M([X,X'], \tau^\perp)$$

$$= g_M(\nabla_X^M X', \tau^\perp) - g_M(\nabla_{X'}^M X, \tau^\perp) - g_M([X,X'], \tau^\perp)$$

$$+ g_M(X', \nabla_X^M \tau^\perp) - g_M(X, \nabla_{X'}^M \tau^\perp)$$

which yields (7.44), since ∇^M is torsion-free. ∎

On the other hand $d\nu = -\kappa^\perp \wedge \nu$ implies $d\kappa^\perp \wedge \nu = 0$, and thus locally $d\kappa^\perp = \gamma \wedge \nu$. This in turn implies $d\kappa^\perp(X,X') = 0$ for $X, X' \in \Gamma L$. As a consequence of (7.44) we have then

$$(7.45) \qquad g_M(X, \nabla_{X'}^M \tau^\perp) = g_M(X', \nabla_X^M \tau^\perp) \quad \text{for } X, X' \in \Gamma L.$$

For the next formula we need the operator $\nabla^M Z : TM \longrightarrow TM$. Note that $g_M(\nabla_Y^M Z, Z) = 0$ for any vectorfield Y, so that in fact $\nabla^M Z : TM \longrightarrow L$. Moreover for $X \in \Gamma L$ we have

$$W(Z)X = -\pi^\perp(\nabla_X^M Z) = -\nabla_X^M Z,$$

so that $\nabla^M Z|L = -W(Z)$. We further claim that

$$(7.46) \qquad \text{trace } ((\nabla^M Z)^2) = \text{trace } (W(Z)^2).$$

For this it suffices to show that $g_M((\nabla^M Z)^2 Z, Z) = 0$. But

$$(\nabla^M Z)^2 Z = \nabla_{\nabla_Z^M Z}^M Z = \nabla_{\tau^\perp}^M Z$$

and $g_M(\nabla^M_{\tau^\perp} Z, Z) = \frac{1}{2} \tau^\perp g_M(Z,Z) = 0$. This proves (7.46). The interest of this formula arises from the selfadjointness of $W(Z) : L \longrightarrow L$, which proves that $W(Z)^2$ is a nonnegative operator, and thus trace $(W(Z)^2) \geq 0$.

For a closed oriented M we are now going to use the following integral formula of Yano [KO, p. 154]

$$(7.47) \quad \int_M \text{Ric}(X,X) \cdot \mu + \int_M \text{trace}((\nabla^M X)^2) \cdot \mu - \int_M (\text{div } X)^2 \cdot \mu = 0,$$

which holds for any vector field X on M, where Ric is the Ricci curvature form on M.

We apply this to the transversal vector field Z of a harmonic foliation $\mathcal{F}$ of codimension one on M. By Theorem 7.35 we have div $Z = 0$, so

$$(7.48) \quad \int_M \text{Ric}(Z,Z) \cdot \mu + \int \text{trace}((\nabla^M Z)^2) \cdot \mu = 0.$$

Next we assume that Ric ≥ 0. Then both integrands in (7.48) are nonnegative. It follows that

$$(7.49) \quad \text{trace}((\nabla^M Z)^2) = \text{trace}(W(Z)^2) = 0.$$

This shows that all eigenvalues of $W(Z)$ are zero. Hence $W(Z) = 0$, and $\mathcal{F}$ is totally geodesic. This proves part of the following result [OS 1][KT 12, 14].

7.50 THEOREM. Let $\mathcal{F}$ be a transversally oriented harmonic foliation of codimension one, on a closed oriented manifold with nonnegative Ricci curvature. Then $\mathcal{F}$ is totally geodesic and Riemannian.

This conclusion has some aspects of the statement of the classical Bernstein Theorem for a single minimal hypersurface in $\mathbb{R}^{n+1}$ (namely if it is complete, and a graph over all of $\mathbb{R}^n$, then for $n \leq 7$ it is necessarily a hyperplane).

PROOF. It remains to show that the harmonicity assumption implies $\tau^\perp = 0$, which characterizes the Riemannian property of $\mathcal{F}$. We first prove

(7.51) $$\text{Ric}(Z,Z) = \text{trace}(\nabla^M \tau^\perp).$$

Let $E_1, \ldots, E_n$ be a local orthonormal frame of L (dim $M = n + 1$). Then Ric (Z,Z) on M^{n+1} is given as an average of sectional curvatures $K(Z,E_i)$ by

(7.52) $$\text{Ric }(Z,Z) = \sum_{i=1}^{n} K(Z,E_i),$$

where

$$K(Z,E_i) = g_M(R(E_i,Z)Z,E_i) = g_M(\nabla^M_{E_i}\nabla^M_Z Z - \nabla^M_Z \nabla^M_{E_i} Z - \nabla^M_{[E_i,Z]} Z, E_i).$$

Now the hypothesis Ric ≥ 0 implies by the preceding arguments that $W(Z) = 0$, hence $\nabla^M_{E_i} Z = 0$ for $i = 1,\ldots,n$. It follows further that

$$[E_i,Z] = \nabla^M_{E_i} Z - \nabla^M_Z E_i = -\nabla^M_Z E_i,$$

and

$$- g_M(\nabla^M_{[E_i,Z]} Z, E_i) = g_M(\nabla^M_{\nabla^M_Z E_i} Z, E_i).$$

Now we expand

$$\nabla^M_Z E_i = \sum_{j=1}^{n} g_M(\nabla^M_Z E_i, E_j) E_j + g_M(\nabla^M_Z E_i, Z) Z,$$

so that

$$g_M(\nabla^M_{\nabla^M_Z E_i} Z, E_i) = \sum_j g_M(\nabla^M_Z E_i, E_j) g_M(\nabla^M_{E_j} Z, E_i) + g_M(\nabla^M_Z E_i, Z) g_M(\nabla^M_Z Z, E_i).$$

Observe that $g_M(\nabla^M_{E_j} Z, E_i) = - g_M(Z, \nabla^M_{E_j} E_i) = 0,$ since we have already proved that $\mathcal{F}$ is totally geodesic, and hence $\nabla^M_{E_j} E_i \in \Gamma L$. Moreover $g_M(\nabla^M_Z E_i, Z) = - g_M(E_i, \nabla^M_Z Z),$ so that

$$g_M(\nabla^M_{\nabla^M_Z E_i} Z, E_i) = - g_M(\tau^\perp, E_i)^2.$$

By (7.52) we have then

$$\text{Ric}(Z,Z) = \sum_{i=1}^{n} g_M(\nabla^N_{E_i} \tau^\perp, E_i) - \sum_{i=1}^{n} g_M(\tau^\perp, E_i)^2.$$

The second sum equals (Theorem of Pythagoras for $\tau^\perp \in \Gamma L$)

$$|\tau^\perp|^2 = g_M(\nabla^M_Z Z, \tau^\perp) = - g_M(Z, \nabla^M_Z \tau^\perp),$$

so that finally

$$\text{Ric }(Z,Z) = \text{trace}(\nabla^M \tau^\perp),$$

as claimed. This formula can equivalently be stated as

(7.53) $$\text{Ric }(Z,Z) = \text{div }\tau^\perp = -\delta\kappa^\perp.$$

The proof above definitely uses the assumption $\text{Ric} \geq 0$, which implies $W(Z) = 0$, and kills many terms in the calculations. The general formula for $\delta\kappa^\perp$ contains additional terms besides $\text{Ric}(Z,Z)$, too numerous to be written out here.

The harmonicity of $\mathcal{F}$ (div $Z = 0$) and $\text{Ric} \geq 0$ imply by (7.48) also $\text{Ric }(Z,Z) = 0$, hence

(7.54) $$\text{div }\tau^\perp = 0.$$

Now we apply (7.47) to $X = \tau^\perp$ and find

(7.55) $$\int_M \text{Ric}(\tau^\perp, \tau^\perp) \cdot \mu + \int_M \text{trace}((\nabla^M \tau^\perp)^2) \cdot \mu = 0.$$

We claim that $\text{trace }((\nabla^M \tau^\perp)^2) \geq 0$ (and hence $= 0$). In fact we prove more precisely

(7.56) $$\text{trace}((\nabla^M \tau^\perp)^2) = \sum_{i,j=1}^n g_M(\nabla^M_{E_i} \tau^\perp, E_j)^2 + |\tau^\perp|^4 \geq 0.$$

Namely

$$\text{trace}((\nabla^M \tau^\perp)^2) = \sum_{i=1}^n g_M((\nabla^M \tau^\perp)^2 E_i, E_i) + g_M((\nabla^M \tau^\perp)^2 Z, Z).$$

Now $(\nabla^M \tau^\perp)^2 E_i = \nabla^M_{\nabla^M_{E_i} \tau^\perp} \tau^\perp$ and $(\nabla^M \tau^\perp)^2 Z = \nabla^M_{\nabla^M_Z \tau^\perp} \tau^\perp$. We expand

$$\nabla^M_{E_i} \tau^\perp = \sum_{j=1}^n g_M(\nabla^M_{E_i} \tau^\perp, E_j) E_j + g_M(\nabla^M_{E_i} \tau^\perp, Z) Z$$

$$\nabla^M_Z \tau^\perp = \sum_{i=1}^n g_M(\nabla^M_Z \tau^\perp, E_i) E_i + g_M(\nabla^M_Z \tau^\perp, Z) Z .$$

Then

$$\text{trace }((\nabla^M \tau^\perp)^2) = \sum_{i,j} g_M(\nabla^M_{E_i} \tau^\perp, E_j) g_M(\nabla^M_{E_j} \tau^\perp, E_i)$$

$$+ \sum_i g_M(\nabla^M_{E_i} \tau^\perp, Z) g_M(\nabla^M_Z \tau^\perp, E_i)$$

$$+ \sum_i g_M(\nabla_Z \tau^\perp, E_i) g_M(\nabla^M_{E_i} \tau^\perp, Z)$$

$$+ g_M(\nabla^M_Z \tau^\perp, Z) g_M(\nabla^M_Z \tau^\perp, Z).$$

The last term equals

$$g_M(\nabla^M_Z \tau^\perp, Z)^2 = (- g_M(\tau^\perp, \nabla^M_Z Z))^2 = (- |\tau^\perp|^2)^2 = |\tau^\perp|^4.$$

The second and third sums are the same. Moreover

$$g_M(\nabla^M_{E_i}\tau^\perp, Z) = - g_M(\tau^\perp, \nabla^M_{E_i} Z) = 0,$$

since $W(Z) = 0$, so these terms vanish. By (7.45) finally

$$\sum_{i,j} g_M(\nabla_{E_i}\tau^\perp, E_j) g_M((\nabla_{E_j}\tau^\perp, E_i) = \sum_{i,j} g_M(\nabla^M_{E_i}\tau^\perp, E_j)^2,$$

which completes the proof of (7.56).

From (7.55) we conclude now $\text{trace}((\nabla^M\tau^\perp)^2) = 0$, and hence in particular $\tau^\perp = 0$ by (7.56). ∎

The situation in Theorem 7.50 is somewhat critical. The conclusion shows that the transversal volume ν is closed as well as coclosed. Thus it is a harmonic 1-form on M. But the existence of harmonic 1-forms is restricted by the Ricci curvature assumptions. If in addition to $\text{Ric} \geq 0$ there is at least a point $x_0 \in M$, such that the Ricci operator at x_0 is positive, then no nontrivial harmonic 1-form exists on M, as follows from the Bochner-Weitzenböck formula [PO, p. 159]

$$-\frac{1}{2}\Delta|\omega|^2 = |\nabla\omega|^2 - g(\Delta\omega,\omega) + \text{Ric}(X,X),$$

valid for a vector field X and its dual 1-form ω. The classical argument of Bochner and Lichnerowicz, with a refinement pointed out in [WU], is as follows. If $\Delta\omega = 0$, one finds by integration

$$\int_M |\nabla\omega|^2 \cdot \mu + \int_M \text{Ric}(X,X) \cdot \mu = 0.$$

The assumption $\text{Ric}(X,X) \geq 0$ implies $\nabla\omega = 0$, so that ω is either nowhere zero or $\omega = 0$. If ω where nowhere zero, then under the strict positivity assumption for the Ricci operator at one point $x_0 \in M$, one would have $\int_M \text{Ric}(X,X) \cdot \mu > 0$, a contradiction.

A situation where these difficulties do not arise is the case of a closed flat manifold. As a consequence one obtains the following result [KT 12].

7.57 COROLLARY. Let $\mathcal{F}$ be a transversally oriented harmonic foliation of codimension one on a closed oriented flat manifold M^{n+1}. Then $\mathcal{F}$ is induced from a hyperplane foliation on the universal covering $\mathbb{R}^{n+1}$.

PROOF. The lift $\tilde{\mathcal{F}}$ to the universal covering $\tilde{M} \cong \mathbb{R}^{n+1}$ is a totally geodesic foliation, hence a foliation by hyperplanes. ∎

CHAPTER 8

FOLIATIONS BY LEVEL HYPERSURFACES

Let $f : M^{n+1} \longrightarrow \mathbb{R}$ be a smooth function. Removing the set of critical points $\mathrm{Crit}(f)$, we obtain a foliation of codimension one on $M - \mathrm{Crit}(f)$ by the level hypersurfaces of f. Let (M, g_M) be Riemannian. The gradient vector field is denoted by the usual ∇f. Then

$$Z = \frac{1}{|\nabla f|} \cdot \nabla f$$

is a unit normal vector field, with dual 1-form

$$\nu = \frac{1}{|\nabla f|} \cdot df.$$

The Hessian of f is the bilinear form on M defined by

(8.1) $$\mathrm{Hess}_f = \nabla^M df \quad \text{i.e.}$$

(8.2) $\quad \mathrm{Hess}_f(X,Y) = (\nabla^M df)(X,Y) = (\nabla^M_X df)(Y) = X(df(Y)) - df(\nabla^M_X Y) = XYf - (\nabla^M_X Y)f.$

Its symmetry is verified by

$$H_f(X,Y) - H_f(Y,X) = [X,Y]f - (\nabla^M_X Y)f + (\nabla^M_Y X)f = 0.$$

Note further, with the notation at the bottom of p. 72 for the isomorphism $\mathcal{V}(M) \longrightarrow \Omega^1(M)$, that

(8.3) $$\nabla^M_X df = \gamma(\nabla^M_X \nabla f),$$

since

$$(\nabla^M_X df)(Y) = X df(Y) - df(\nabla^M_X Y) = X g_M(\nabla f, Y) - g_M(\nabla f, \nabla^M_X Y) = g_M(\nabla^M_X \nabla f, Y).$$

Next we show that for $X, X' \in \Gamma L$ (tangent to the foliation defined by f)

(8.4) $$\text{Hess}_f(X, X') = -|\nabla f| \cdot g_M(W(Z)X, X').$$

PROOF. Let $\lambda = |\nabla f|$. By (8.1)(8.3)

$$\text{Hess}_f(X, X') = g_M(\nabla^M_X \nabla f, X') = g_M(X\lambda \cdot Z + \lambda \nabla^M_X Z, X')$$

$$= \lambda g_M(\nabla^M_X Z, X') = \lambda g_M(\pi^\perp(\nabla^M_X Z), X') = -\lambda g_M(W(Z)X, X'). \blacksquare$$

From this we find for the second fundamental form $\alpha : L \otimes L \longrightarrow Q$, as defined in Chapter 6,

$$g_M(\alpha(X, X'), Z) = g_M(W(Z)X, X') = -\frac{1}{|\nabla f|} \cdot \text{Hess}_f(X, X'),$$

and it follows that

(8.5) $$\alpha(X, X') = -\frac{1}{|\nabla f|} \cdot \text{Hess}_f(X, X') \cdot Z.$$

Next we calculate the mean curvature

$$\kappa(Z) = \text{trace } W(Z) = \sum_{i=1}^{n} g_M(W(Z)E_i, E_i)$$

for an orthonormal frame $E_1, \ldots, E_n$ of L. By (8.4) we find

(8.6) $$\kappa(Z) = -\frac{1}{|\nabla f|} \cdot \sum_{i=1}^{n} \text{Hess}_f(E_i, E_i),$$

(sum up to n only). Now for an orthonormal frame $E_1, \ldots, E_{n+1}$ of TM

$$\Delta f = \delta df = -\sum_{A=1}^{n+1} (\nabla_{E_A} df)(E_A) = -\sum_{A=1}^{n+1} (E_A(df(E_A)) - df(\nabla^M_{E_A} E_A)),$$

which shows $\Delta f = -\text{trace Hess}_f$. From (8.6) we find therefore

(8.7) $$\kappa(Z) = \frac{1}{|\nabla f|} \cdot (\Delta f + \text{Hess}_f(Z,Z)).$$

Note that the sign convention for $\Delta f = \delta df$ is such that on $\mathbb{R}^{n+1}$ we have $\Delta f = -\sum_{A=1}^{n+1} \frac{\partial^2 f}{\partial x_A^2}$. Further $\nabla^M_Z Z \in \Gamma L$, hence in (8.7)

$$\text{Hess}_f(Z,Z) = ZZf - (\nabla^M_Z Z)f = ZZf.$$

Note further, that with $\lambda = |\nabla f|$

$$\text{Hess}_f(Z,Z) = (\nabla^M_Z df)(Z) = g_M(\nabla^M_Z \nabla f, Z) = \frac{1}{\lambda^2} g_M(\nabla^M_{\nabla f} \nabla f, \nabla f) = \frac{1}{\lambda^2} \cdot \frac{1}{2} \nabla f g_M(\nabla f, \nabla f) =$$

$$\frac{1}{2\lambda^2} \nabla f(\lambda^2) = \frac{1}{\lambda} \nabla f(\lambda) = \frac{1}{\lambda} d\lambda(\nabla f) = \frac{1}{\lambda} g_M(\nabla f, \nabla \lambda).$$

It follows that (8.7) can be written equivalently as

(8.8) $\quad \kappa(Z) = \frac{1}{|\nabla f|} [\Delta f + ZZf] = \frac{1}{|\nabla f|} [\Delta f + \frac{1}{|\nabla f|} \cdot g_M(\nabla f, \nabla(|\nabla f|))]$.

In the case of a harmonic function f this leads to a generally nontrivial mean curvature $\kappa(Z) = \frac{1}{|\nabla f|} \cdot ZZf$, thus not generally to a harmonic foliation in the sense of Chapters 5 and 6.

We verify that for the foliation on $\mathbb{R}^{n+1} - \{0\}$ of Example 7.12 defined by the level hypersurfaces of the function $r = (\Sigma x_i^2)^{1/2}$, these formulas lead to the value of the mean curvature previously calculated in (7.20). Since $\Delta r = \delta dr = - \text{div} \nabla r$, and $\nabla r = \frac{1}{r} \sum_i x_i \frac{\partial}{\partial x_i}$, it follows $\Delta r = -\frac{n}{r}$. Further $\text{Hess}_r(Z,Z) = ZZr$, since $\nabla_Z Z = 0$. But $Zr = 1$, hence $ZZr = 0$, and (8.7) reduces to

$$\kappa(Z) = \Delta r = -\frac{n}{r},$$

coinciding with (7.20). In this example $\lambda = |\nabla r| = 1$ and $\nabla \lambda = 0$, so that the correction term in (8.8) disappears.

8.9 THEOREM. Let $\mathcal{F}$ be the foliation on M - Crit(f) defined by the level hypersurfaces of $f : M \longrightarrow \mathbb{R}$. Let g_M be a Riemannian metric on M. Then the following conditions are equivalent:
  (i) $\mathcal{F}$ is Riemannian and g_M bundle-like;
  (ii) $X|\nabla f|^2 = 0$ for all $X \in \Gamma L$;
  (iii) $\text{Hess}_f(X, \nabla f) = 0$ for all $X \in \Gamma L$;
  (iv) $[X,Z]f = 0$ for all $X \in \Gamma L$.

PROOF. First we calculate for any vector field Y by (8.1) (8.3)

$$(8.10) \quad \text{Hess}_f(Y, \nabla f) = (\nabla^M_Y df)(\nabla f) = g_M(\nabla^M_Y \nabla f, \nabla f) = \tfrac{1}{2} Y g_M(\nabla f, \nabla f) = \tfrac{1}{2} Y |\nabla f|^2.$$

If we restrict to $X \in \Gamma L$, this formula shows the equivalence of (ii) and (iii).

Next we consider the mean curvature vector field $\tau^\perp = \nabla^M_Z Z \in \Gamma L$ of $\mathcal{F}^\perp$, and find with $\lambda = |\nabla f|$

$$\tau^\perp = \nabla^M_Z (\tfrac{1}{\lambda} \cdot \nabla f) = Z(\tfrac{1}{\lambda}) \cdot \nabla f + \tfrac{1}{\lambda} \cdot \nabla^M_Z \nabla f.$$

For $X \in \Gamma L$ this yields by (8.1) (8.3) (8.10)

$$(8.11) \quad g_M(\tau^\perp, X) = \tfrac{1}{\lambda} g_M(\nabla^M_Z \nabla f, X) = \tfrac{1}{\lambda} \text{Hess}_f(Z, X) = \tfrac{1}{\lambda^2} \text{Hess}_f(\nabla f, X)$$

$$= \tfrac{1}{2\lambda^2} X(\lambda^2).$$

Since $\tau^\perp \in \Gamma L$, this expression vanishes for all $X \in \Gamma L$ iff $\tau^\perp = 0$, i.e. iff $\mathcal{F}$ is Riemannian (Theorem 7.3). This proves the equivalence of (i) and (ii).

We have further

$$g_M(\tau^\perp, X) = g_M(\nabla^M_Z Z, X) = - g_M(Z, \nabla^M_Z X) = - g_M(Z, \nabla^M_X Z + [Z, X]) = g_M(Z, [X, Z])$$

which proves the equivalence of (i) and (iv). ∎

It is instructive to calculate $d\nu$, the vanishing of which is by Theorem 7.3 equivalent to the Riemannian property. For $\nu = \tfrac{1}{|\nabla f|} df$ we have

$$d\nu = d(\tfrac{1}{|\nabla f|}) \wedge df.$$

Since $d\nu(X,X')$ clearly vanishes for $X,X' \in \Gamma L$, the vanishing of $d\nu$ is equivalent to the vanishing of $i(X)i(Z)d\nu$ for all $X \in \Gamma L$. But

$$d\nu(Z,X) = Z(\tfrac{1}{|\nabla f|}) \cdot Xf - X(\tfrac{1}{|\nabla f|}) \cdot Zf.$$

The first term on the RHS vanishes, and

$$Zf = \tfrac{1}{|\nabla f|} \cdot (\nabla f)f = \tfrac{1}{|\nabla f|} \cdot df(\nabla f) = \tfrac{1}{|\nabla f|} \cdot g_M(\nabla f, \nabla f) = |\nabla f|.$$

Thus

$$(8.12) \qquad d\nu(Z,X) = \tfrac{1}{|\nabla f|} \cdot X|\nabla f| = \frac{1}{2|\nabla f|^2} \cdot X|\nabla f|^2,$$

and the vanishing of this expression is condition (ii) in Theorem 8.9.

Next we characterize foliations by level hypersurfaces which are harmonic.

8.13 THEOREM. <u>Let $\mathcal{F}$ be the foliation on $M - \mathrm{Crit}(f)$ defined by the level hypersurfaces of</u> $f : M \longrightarrow \mathbb{R}$. <u>Let g_M be a Riemannian metric on</u> M. <u>Then the following conditions are equivalent</u>:

(i) $\mathcal{F}$ <u>is harmonic (i.e. all leaves are minimal submanifolds)</u>;
(ii) $\Delta f + \dfrac{1}{|\nabla f|^2} \cdot \mathrm{Hess}_f(\nabla f, \nabla f) = 0;$
(iii) $\Delta f + \dfrac{1}{|\nabla f|} \cdot g_M(\nabla f, \nabla f(|\nabla f|)) = 0;$
(iv) $\mathrm{div}\,(\tfrac{1}{|\nabla f|} \cdot \nabla f) = 0.$

PROOF. (i) ⇔ (ii) ⇔ (iii) follows from (7.63) (7.64). (i) ⇔ (iv) follows from Theorem 7.35. ■

Note that generally $\operatorname{div}(hX) = Xh + h \cdot \operatorname{div} X$, so that

$$\operatorname{div}\left[\frac{1}{|\nabla f|} \cdot \nabla f\right] = \nabla f\left[\frac{1}{|\nabla f|}\right] + \frac{1}{|\nabla f|} \cdot \operatorname{div} \nabla f.$$

With our sign convention $\operatorname{div} \nabla f = -\delta\, df = -\Delta f$, so that

(8.14)
$$\operatorname{div}\left[\frac{1}{|\nabla f|} \cdot \nabla f\right] = \nabla f\left[\frac{1}{|\nabla f|}\right] - \frac{1}{|\nabla f|} \cdot \Delta f$$

$$= -\frac{1}{|\nabla f|^2} \cdot \nabla f(|\nabla f|) - \frac{1}{|\nabla f|} \cdot \Delta f$$

$$= -\frac{1}{|\nabla f|^2} \cdot g_M(\nabla(|\nabla f|), \nabla f) - \frac{1}{|\nabla f|} \cdot \Delta f$$

which proves (again) the equivalence of condition (iii) and (iv).

Comparing the last calculation with (8.8) shows that more generally

$$\kappa(Z) = -\operatorname{div}\left[\frac{\nabla f}{|\nabla f|}\right].$$

Since $Z = \frac{\nabla f}{|\nabla f|}$, this follows also from (7.11).

The case of a totally geodesic foliation $\mathcal{F}$ corresponds by formula (8.5) to the condition

$$\operatorname{Hess}_f(X, X') = 0 \quad \text{for all} \quad X, X' \in \Gamma L.$$

An interesting case to consider is the case of a harmonic function $f : M \longrightarrow \mathbb{R}$. The mean curvature is by (8.8)

(8.15) $$\kappa(Z) = \frac{1}{|\nabla f|^2} \cdot g_M(\nabla f, \nabla(|\nabla f|)).$$

Note that the harmonicity of the foliation by level hypersurfaces is characterized by $\kappa = 0$, which is not necessarily the case for the level surfaces of a harmonic function.

Note further that (still for harmonic f) the Riemannian property is characterized by $\text{Hess}_f(X,\nabla f) = 0$, for all X tangent to $\mathcal{F}$, while the minimality of the level surfaces is characterized by $\text{Hess}_f(\nabla f, \nabla f) = 0$ (∇f is normal to $\mathcal{F}$). These properties are both satisfied, if $\text{Hess}_f(Y,\nabla f) = 0$ for all vector fields Y, i.e. iff ∇f is in the null space of Hess_f. By (8.10) this means precisely that $Y|\nabla f|^2 = 0$ for all $Y \in \Gamma TM$, or $|\nabla f|$ = constant. We summarize these facts, which are of interest only if M is noncompact.

8.16 COROLLARY. Let $\mathcal{F}$ be the foliation on $M - \text{Crit}(\mathcal{F})$ defined by the level hypersurfaces of a harmonic function $f : M \longrightarrow \mathbb{R}$ on the Riemannian manifold (M, g_M). Then the following holds.
 (i) $\mathcal{F}$ is Riemannian iff $\text{Hess}_f(X,\nabla f) = 0$ for all $X \in \Gamma L$.
 (ii) $\mathcal{F}$ has all leaves minimal iff $\text{Hess}_f(\nabla f, \nabla f) = 0$.
 (iii) $\mathcal{F}$ has both properties iff $|\nabla f|$ is constant on M.

8.17 EXAMPLE. We consider a domain $D \subset \mathbb{C}$ and a holomorphic function $f : D \longrightarrow \mathbb{C}$. Let $f = u + iv$, with $u,v : D \longrightarrow \mathbb{R}$. The Cauchy-Riemann equations for f are

$$u_x = v_y, \ u_y = -v_x,$$

and u and v are harmonic functions. Further

$$\nabla u \cdot \nabla v = u_x v_x + u_y v_y = 0$$

so u,v define two orthogonal foliations $\mathcal{F}$ and $\mathcal{F}^\perp$ by level curves on D with the critical points removed. The (mean) curvature of the foliation $\mathcal{F}$ is then given by

$$\kappa(Z) = \frac{1}{|\nabla u|^2} \cdot \langle \nabla u, \nabla(|\nabla u|) \rangle.$$

Now $\nabla u = u_x \frac{\partial}{\partial x} + u_y \frac{\partial}{\partial y}$ and $|\nabla u|^2 = u_x^2 + u_y^2$. Further

$$\nabla |\nabla u|^2 = 2|\nabla u| \cdot \nabla |\nabla u|,$$

so that

$$\nabla |\nabla u|^2 = \nabla(u_x^2 + u_y^2) = 2u_x(u_{xx}\frac{\partial}{\partial x} + u_{xy}\frac{\partial}{\partial y}) + 2u_y(u_{yx}\frac{\partial}{\partial x} + u_{yy}\frac{\partial}{\partial y}),$$

yields

$$\nabla |\nabla u| = \frac{1}{2|\nabla u|} \cdot \nabla |\nabla u|^2.$$

It follows that

$$\kappa(Z) = \frac{1}{2|\nabla u|^3} [u_x(2u_x u_{xx} + 2u_y u_{yx}) + u_y(2u_x u_{xy} + 2u_y u_{yy})]$$

$$= \frac{1}{(u_x^2 + u_y^2)^{3/2}} [u_x^2 \cdot u_{xx} + 2u_x u_y u_{xy} + u_y^2 \cdot u_{yy}].$$

To compare this with the usual formula note that in this calculation u is assumed to be harmonic, i.e. $u_{xx} + u_{yy} = 0$, and that the sign of the curvature depends on the orientation convention (see Jerrard and Rubel [JR] and [J] for further comments on this situation).

The differentials of conjugate harmonic functions u,v, arising from a holomorphic function $f = u + iv$, are related as follows. Note that in $\mathbb{R}^2$

$$*dx = dy \quad \text{and} \quad *dy = -dx$$

so that

$$*du = *(u_x dx + u_y dy) = u_x dy - u_y dx = v_x dx + v_y dy = dv.$$

While $\frac{1}{|\nabla u|} \cdot du$ is the transversal volume for the foliation by the level curves of u, the form $\frac{1}{|\nabla u|} dv$ is its characteristic form.

The condition $|\nabla f|$ = constant is the eiconal property of optics. It is satisfied for the foliation of $\mathbb{R}^{n+1} - \{0\}$ by concentric spheres, the level hypersurfaces of the function r with $|dr| = 1$.

The eiconal property is in particular satisfied for a function f with vanishing Hessian. Since $\Delta f = -\text{trace Hess}_f$, such a function is necessarily harmonic, and the foliation defined by f is Riemannian as well as harmonic. By (8.1) (8.3) the Hessian vanishes precisely when ∇f is a parallel vectorfield on M. By (8.10) it follows then that $|\nabla f|$ = constant on M. For the mean curvature of the level hypersurfaces of f it follows by (8.7) that $\kappa = 0$.

Note that for any curve $\gamma : I \longrightarrow M$ and any function $f : M \longrightarrow \mathbb{R}$ we have

$$f(\gamma(t))' = \gamma'f = g_M(\nabla f, \gamma')$$

and

$$f(\gamma(t))'' = \gamma' g_M(\nabla f, \gamma') = g_M(\nabla^M_{\gamma'}\nabla f, \gamma') + g_M(\nabla f, \nabla^M_{\gamma'}\gamma')$$

$$= \text{Hess}_f(\gamma', \gamma') + g_M(\nabla f, \nabla^M_{\gamma'}\gamma').$$

For a geodesic γ with arc length $t = s$ we have $\nabla^M_{\dot\gamma}\dot\gamma = 0$ (notation $\frac{d}{ds} = \cdot$), and therefore

(8.18) $$f(\gamma)^{\cdot\cdot} = \text{Hess}_f(\dot\gamma, \dot\gamma).$$

From the previous discussion we obtain the following result.

8.19 PROPOSITION. <u>For a function</u> $f : M \longrightarrow \mathbb{R}$ <u>the following conditions are equivalent</u>:
 (i) ∇f <u>is a parallel vector field</u>;
 (ii) $\text{Hess}_f = 0$;
 (iii) $f(\gamma)^{\cdot\cdot} = 0$ <u>for any geodesic</u> $\gamma(s)$ <u>parametrized by arc length</u> s.

Next we examine the behaviour of the mean curvature along the leaves in the Riemannian case. First some preparations.

8.20 LEMMA. <u>Let</u> $X|\nabla f|^2 = 0$ <u>for all</u> $X \in \Gamma L$. <u>Then</u> $[X, \nabla f] \in \Gamma L$.

PROOF. By Theorem 8.9, part (iv) we know that $[X, Z] \in \Gamma L$. But

$$[X, \tfrac{1}{|\nabla f|} \cdot \nabla f] = X\!\left[\tfrac{1}{|\nabla f|}\right] \cdot \nabla f + \tfrac{1}{|\nabla f|} \cdot [X, \nabla f].$$

Since the first term on the RHS vanishes if $X|\nabla f|^2 = 0$, the desired result follows. ∎

8.21 LEMMA. Let $X|\nabla f|^2 = 0$ for all $X \in \Gamma L$. Then

$$X(\frac{1}{|\nabla f|^2} \cdot g_M(\nabla f, \nabla(|\nabla f|))) = 0.$$

PROOF. The LHS equals

$$\frac{1}{|\nabla f|^2} \cdot Xg_M(\nabla f, \nabla(|\nabla f|)) = \frac{1}{2|\nabla f|^3} \cdot Xg_M(\nabla f, 2 \cdot |\nabla f| \cdot \nabla|\nabla f|)$$

$$= \frac{1}{2|\nabla f|^3} \cdot Xg_M(\nabla f, \nabla|\nabla f|^2) = \frac{1}{2|\nabla f|^3} \cdot X(d(|\nabla f|^2)(\nabla f))$$

$$= \frac{1}{2|\nabla f|^3} \cdot (X\nabla f)(|\nabla f|^2)$$

$$= \frac{1}{2|\nabla f|^3} \cdot [X, \nabla f](|\nabla f|^2) + (\nabla f)X(|\nabla f|^2).$$

The second term vanishes, since by the assumption $X|\nabla f|^2 = 0$. The first term vanishes by Lemma 8.20. ∎

8.22 THEOREM. Let $f : M \longrightarrow \mathbb{R}$ be a function, such that on $M - \text{Crit}(f)$ we have $X|\nabla f|^2 = 0$ for all X tangent to the hypersurfaces $f = \text{constant}$. Then for the derivative of the mean curvature $\kappa(Z)$ of these surfaces we have

(8.23) $$X\kappa(Z) = \frac{1}{|\nabla f|} \cdot X(\Delta f).$$

PROOF. This follows from (8.8) and Lemma 8.21. ∎

An isoparametric function on a Riemannian manifold is a function $f : M \longrightarrow \mathbb{R}$ satisfying on $M - \text{Crit}(f)$ the system of differential equations

$$X|\nabla f|^2 = 0$$

$$X(\Delta f) = 0$$

for all vector fields X tangent to the foliation defined by f (a concept originally introduced by Elie Cartan). For such a function, it follows from Theorem 8.22 that each level hypersurface of f has constant mean curvature. This fact is of particular interest for space forms. In that case one can conclude from this that all eigenvalues of the Weingarten map are constants on each level hypersurface of f, which leads to beautiful structure theorems (see e.g. [MZ 1,2] and [NZ]).

CHAPTER 9
INFINITESIMAL AUTOMORPHISMS AND BASIC FORMS

Let $\mathcal{F}$ be an arbitrary foliation on M (no metric yet to be involved). A vector field $Y \in \Gamma TM$ is an infinitesimal automorphism of $\mathcal{F}$, if

(9.1) $\qquad [X,Y] \in \Gamma L \quad \text{for all} \quad X \in \Gamma L.$

This means that the flow of Y preserves the foliation, i.e. maps leaves into leaves. The Lie algebra ΓL acts on ΓQ by

(9.2) $\qquad \theta(X)s = \pi[X,Y_s] = \overset{\circ}{\nabla}_X s \quad \text{for} \quad X \in \Gamma L,\ s \in \Gamma Q,$

where $Y_s \in \Gamma TM$ with $\pi(Y_s) = s$. In terms of this action (9.1) can be written

(9.3) $\qquad \theta(X)Y = \pi[X,Y] = \overset{\circ}{\nabla}_X Y = 0$

for $X \in \Gamma L$, $Y = \pi(Y)$. In a distinguished chart $(x_1, \ldots, x_p;\ y_1, \ldots, y_q)$ such a vector field is of the form

$$Y = \sum_{i=1}^{p} a_i \frac{\partial}{\partial x_i} + \sum_{\alpha=1}^{q} b_\alpha \frac{\partial}{\partial y_\alpha}$$

with $a_i = a_i(x,y)$ and $\frac{\partial}{\partial x_i} b_\alpha = 0$, i.e. $b_\alpha = b_\alpha(y)$. Then

(9.4) $\qquad Y = \sum_{\alpha=1}^{q} b_\alpha \frac{\partial}{\partial y_\alpha}.$

Such a vector field is called projectable. Molino [M 7,8] calls such a Y foliated vector field and Y its associated transverse field. $V(\mathcal{F})$ denotes the set of all infinitesimal automorphisms of $\mathcal{F}$. It is a Lie algebra under the natural bracket. Let further ΓQ^L denote the invariant sections of Q under the action (9.2). The exact bundle sequence $0 \longrightarrow L \longrightarrow TM \longrightarrow Q \longrightarrow 0$ induces an exact sequence of vectorspaces

(9.5) $$0 \longrightarrow \Gamma L \longrightarrow V(\mathcal{F}) \longrightarrow \Gamma Q^L \longrightarrow 0$$

$$Y \longmapsto Y$$

This is an exact sequence of Lie algebras. We finally note that in a distinguished chart there exist local framings of Q by transverse fields, namely $\frac{\partial}{\partial y_1}, \ldots, \frac{\partial}{\partial y_q}$. In the presence of a metric, they can be lifted to give a local framing of $L^\perp$ by infinitesimal automorphisms.

BASIC FORMS. Let $\mathcal{F}$ be an arbitrary foliation. A differential form $\omega \in \Omega^r(M)$ is basic, if

(9.6) $$i(X)\omega = 0, \; \theta(X)\omega = 0 \quad \text{for} \quad X \in \Gamma L.$$

In a distinguished chart $(x_1, \ldots, x_p; y_1, \ldots, y_q)$ of $\mathcal{F}$ this means that

$$\omega = \sum_{\alpha_1 < \ldots < \alpha_r} \omega_{\alpha_1 \ldots \alpha_r} dy_{\alpha_1} \wedge \ldots \wedge dy_{\alpha_r}$$

where the functions $\omega_{\alpha_1 \ldots \alpha_r}(y)$ are independent of x, i.e. $\frac{\partial}{\partial x_i} \omega_{\alpha_1 \ldots \alpha_r} = 0$. The exterior derivative preserves basic forms, since

$\theta(X)d\omega = d\theta(X)\omega = 0$, $i(X)d\omega = \theta(X)\omega - di(X)\omega = 0$ for ω basic. Thus the set $\Omega_B^\cdot \equiv \Omega_B^\cdot(\mathcal{F})$ of all basic forms constitutes a subcomplex

$$d : \Omega_B^r \longrightarrow \Omega_B^{r+1}$$

of the De Rham complex $\Omega^\cdot(M)$. We also denote $d|\Omega_B = d_B$. Its cohomology

(9.7) $$H_B^\cdot \equiv H_B^\cdot(\mathcal{F}) = H(\Omega_B^\cdot(\mathcal{F}), d_B)$$

is the basic cohomology of $\mathcal{F}$. It plays the role of the De Rham cohomology of the leaf space of the foliation.

For the case of codimension one discussed in Chapters 7 and 8, there are just two groups $H_B^0(\mathcal{F})$ and $H_B^1(\mathcal{F})$. In the general case we have the following facts.

Let $r = 0$ (and M connected). Then the 0-cycles $Z_B^0(\mathcal{F})$ are functions $f : M \longrightarrow \mathbb{R}$ which are locally constant, hence constant. Thus

(9.8) $$H_B^0(\mathcal{F}) \cong \mathbb{R}.$$

For $r = 1$ the situation is as follows.

9.9 PROPOSITION. <u>The inclusion</u> $\Omega_B^\cdot(\mathcal{F}) \longrightarrow \Omega^\cdot(M)$ <u>induces an injective map</u>

$$H_B^1(\mathcal{F}) \longrightarrow H_{DR}^1(M).$$

PROOF. Let $\omega \in \Omega_B^1(\mathcal{F})$ such that $\omega = df$ for $f : M \longrightarrow \mathbb{R}$. For $X \in \Gamma L$ we have

$$Xf = i(X)df = i(X)\omega = 0.$$

It follows that $f \in \Omega_B^0(\mathcal{F})$ and $[\omega] = 0 \in H_B^1(\mathcal{F})$. ∎

It was shown by Schwarz [SCH] that nothing of the sort is true for $r \geq 2$. There exist foliations with $H_B^r(\mathcal{F})$ infinite-dimensional for $2 \leq r \leq q$. His examples are smooth. Ghys [GH 4] has constructed for each $q \geq 2$ a real analytic flow on a closed $(q + 1)$-manifold such that the cohomology spaces $H_B^r(\mathcal{F})$ are of uncountable infinite dimension for $2 \leq r \leq q$.

SPECTRAL SEQUENCE of $\mathcal{F}$. This is the spectral sequence determined by the following multiplicative filtration of the De Rham complex $\Omega^{\cdot} = \Omega^{\cdot}(M)$ [KT 1, 2, 3, 11]

$$(9.10) \quad F^r\Omega^m = \{\omega \in \Omega^m \mid i(X_1) \ldots i(X_{m-r+1})\omega = 0 \text{ for } X_1, \ldots, X_{m-r+1} \in \Gamma L\}.$$

It is a convenient tool to formulate many properties of foliations. It is a decreasing filtration by differential ideals. Clearly

$$F^0\Omega^m = \Omega^m \quad \text{and} \quad F^{m+1}\Omega^m = 0.$$

Further

$$(9.11) \quad F^r\Omega^{p+r} = \Omega^{p+r} \quad (p = \dim \mathcal{F}),$$

since $(p + r) - r + 1 = p + 1$, and every $(p + r)$-form evaluated on $p + 1$ vector fields tangent to $\mathcal{F}$ vanishes. Note that $F^{r+1}\Omega^{p+r}$ consists of all

$(p + r)$-forms evaluating to zero on $(p + r) - (r + 1) + 1 = p$ vector fields tangent to $\mathcal{F}$. Thus by definition

(9.12) $\qquad F^{r+1}\Omega^{p+r} \equiv$ "$\mathcal{F}$-trivial" $(p + r)$-forms.

Rummler's formula (6.17), for a foliation $\mathcal{F}$ on a Riemannian manifold, can be restated as

(9.13) $\qquad \theta(Z)\chi_{\mathcal{F}} + \kappa(Z) \cdot \chi_{\mathcal{F}} \in F^1\Omega^p,$

or equivalently

(9.14) $\qquad d\chi_{\mathcal{F}} + \kappa \wedge \chi_{\mathcal{F}} \equiv \varphi_0 \in F^2\Omega^{p+1}.$

An equivalent description of the filtration (9.10) is as forms $\omega \in \Omega^m$, which are locally sums of products

(9.15) $\qquad \alpha \wedge \beta$ with $\alpha \in \wedge^r \mathbf{Q}^*, \quad r \leq m$

and β of degree $m - r$. Clearly for such a form $i(X_1) \ldots i(X_{m-r+1})\omega = 0$, and conversely every $\omega \in F^r\Omega^m$ has locally a representation as a sum of forms as in (9.15). Since $\wedge^{q+1}\mathbf{Q}^* = 0$, this shows that

(9.16) $\qquad F^{q+1}\Omega^m = 0 \quad (q = \mathrm{codim}\ \mathcal{F}).$

Note that for $q = 1$ this condition implies in particular

$$F^2\Omega^{p+1} \equiv \mathcal{F}\text{-trivial } (p + 1)\text{-forms} = 0$$

which is the reason why in this case the form φ_0 in (9.14) vanishes, a fact heavily exploited in Chapter 7.

The initial term $E_0^{r,s}$ of the spectral sequence is defined as the quotient in the exact sequence

(9.17) $\qquad 0 \longrightarrow F^{r+1}\Omega^{r+s} \longrightarrow F^r\Omega^{r+s} \longrightarrow G^r\Omega^{r+s} \equiv E_0^{r,s} \longrightarrow 0.$

In the presence of a metric g_M this sequence is split, and an element $\bar{\omega} \in E_0^{r,s}$ is represented by an $(r + s)$-form ω which is locally a sum of forms

(9.18) $\qquad\qquad \alpha \wedge \beta \quad \text{with} \quad \alpha \in \wedge^r Q^*, \; \beta \in \wedge^s L^*.$

The differential d_0 of bidegree $(0,1)$ in

$$E_0^{r,s} \cong \text{Hom}(\wedge^s L, \wedge^r Q^*)$$

corresponds to the Chevalley-Eilenberg differential d_C on the RHS. Its homology

$$E_1^{r,s} = H^s_{d_C}(\text{Hom}(\wedge^\cdot L, \wedge^r Q^*))$$

gives for $s = 0$

$$E_1^{r,0} = \Gamma((\wedge^r Q^*)^L) \cong \Omega_B^r(\mathcal{F}).$$

The differential d_1 of bidegree $(1,0)$ induces on $\Omega_B^\cdot(\mathcal{F})$ precisely the differential d_B, so that

$$E_2^{r,0} \cong H_B^r(\mathcal{F}).$$

The basic cohomology of $\mathcal{F}$ is thus the basis of the E_2-term of the spectral sequence associated to $\mathcal{F}$. For the case of a fibration, this corresponds to the position of the De Rham cohomology of the basis in the spectral sequence associated to the fibration.

INFINITESIMAL AUTOMORPHISMS AGAIN. An infinitesimal automorphism $Y \in V(\mathcal{F})$ acts on $s \in \Gamma Q$ by

(9.19) $$\theta(Y)s = \pi[Y, Y_s],$$

where $Y_s \in \Gamma TM$ such that $\pi(Y_s) = s$. Because $Y \in V(\mathcal{F})$, the RHS does not depend on the choice of Y_s. For the particular case of $Y = X \in \Gamma L$, this reduces to (9.2).

Let more generally ω be any covariant tensor of degree r on Q. Then an infinitesimal automorphism Y acts on ω by

(9.20) $$(\theta(Y)\omega)(s_1, \ldots, s_r) = Y\omega(s_1, \ldots, s_r) - \sum_{i=1}^{r} \omega(s_1, \ldots, \theta(Y)s_i, \ldots, s_r).$$

For $Y = X \in \Gamma L$ this reduces to the action considered in Chapter 5.

Let now $\omega \in \Omega_B^r(\mathcal{F})$ and $Y \in V(\mathcal{F})$. Then the formulas

$$i(X)\theta(Y)\omega = \theta(Y)i(X)\omega - i[X,Y]\omega$$

$$\theta(X)\theta(Y)\omega = \theta(Y)\theta(X)\omega + \theta[X,Y]\omega$$

show that $\theta(Y)\omega$ is again a basic form. Similary $i(Y)\omega$ is also a basic form. Thus we have $V(\mathcal{F})$ acting on $\Omega_B(\mathcal{F})$, via $i(Y)$ and $\theta(Y)$ with the usual relations. This is just one aspect of the fact that infinitesimal automorphisms are compatible with the filtration (9.10).

HARMONIC FOLIATIONS. Consider a transversally oriented foliation $\mathcal{F}$ of codimension q on a Riemannian manifold (M, g_M), with a holonomy invariant transverse volume ν. Since $i(X)\nu = 0$ for $X \in \Gamma L$, the holonomy invariance condition $\theta(X)\nu = 0$ for $X \in \Gamma L$ shows that $\nu \in \Omega_B^q(\mathcal{F})$. It follows that $d\nu = 0$. It is of interest to examine the cohomology class $[\nu] \in H_B^q(\mathcal{F})$, which plays the role of the orientation class for the leaf space of $\mathcal{F}$. Of particular importance are conditions implying the nontriviality of this class.

9.21 THEOREM. Let $\mathcal{F}$ be a transversally oriented harmonic foliation of codimension q, with invariant transversal volume ν on a closed oriented Riemannian manifold (M, g_M). Then $[\nu] \neq 0$ in $H_B^q(\mathcal{F})$.

PROOF. Assume $\nu = d\alpha$ with $\alpha \in \Omega_B^{q-1}$, and let $\chi_{\mathcal{F}} = *\nu$ be the characteristic form of $\mathcal{F}$. Then

$$d(\alpha \wedge \chi_{\mathcal{F}}) = d\alpha \wedge \chi_{\mathcal{F}} + (-1)^{q-1} \alpha \wedge d\chi_{\mathcal{F}}.$$

Since $\alpha \in \Omega_B^{q-1}$, it follows that $\alpha \in F^{q-1}$ (for simplicity we write $F^{q-1} = F^{q-1}\Omega^{q-1}$). The harmonicity assumption on $\mathcal{F}$ implies by (9.14) that $d\chi_{\mathcal{F}} \in F^2$. It follows that the second term on the RHS is of filtration degree $q - 1 + 2 = q + 1$, and hence vanishes by (9.16). Thus

$$d(\alpha \wedge \chi_{\mathcal{F}}) = d\alpha \wedge \chi_{\mathcal{F}} = \nu \wedge \chi_{\mathcal{F}} = \mu = \text{volume form of } g_M.$$

This contradicts the fact that $[\mu] \neq 0$ in $H^n(M)$. ∎

9.22 COROLLARY [KT 9]. Let $\mathcal{F}$ be a transversally oriented and taut Riemannian foliation of codimension q on a closed oriented manifold M^n. Then $H_B^q(\mathcal{F}) \neq 0$.

PROOF. Recall that $\mathcal{F}$ is taut if there exists a metric on M such that all the leaves of $\mathcal{F}$ are minimal submanifolds. For a Riemannian foliation, this can be achieved with a bundle-like metric [KT 6, Cor. 2.31]. The corresponding transversal volume ν is then holonomy invariant, and Theorem 9.21 applies to prove $[\nu] \neq 0$. It follows that $H_B^q(\mathcal{F}) \neq 0$. ∎

Another application concerns (transversally) symplectic foliations [D][KT 4,11]. For such a foliation $\mathcal{F}$ of codimension $q = 2m$ on M^n there exists a basic and closed 2-form $\omega \in \Omega_B^2(\mathcal{F})$ such that ω^m is a nowhere zero q-form. Note that ω^k for $k = 1, \ldots, m$ is closed and satisfies

$$i(X)\omega^k = \sum \omega \wedge \ldots \wedge i(X)\omega \wedge \ldots \wedge \omega = 0$$

$$\theta(X)\omega^k = \sum \omega \wedge \ldots \wedge \theta(X)\omega \wedge \ldots \wedge \omega = 0.$$

Thus $\omega^k \in \Omega_B^{2k}(\mathcal{F})$ give rise to basic cohomology classes $[\omega^k] \in H_B^{2k}(\mathcal{F})$.

9.23 THEOREM. Let $\mathcal{F}$ be a taut and (transversally) symplectic foliation of codimension $q = 2m$ on a closed oriented manifold M^n. Then the cohomology classes $[\omega^k] \in H_B^{2k}(\mathcal{F})$ for $k = 1, \ldots, m$ are all nontrivial.

PROOF. Let g_M be a metric such that all leaves are minimal. The q-form $\tilde{\nu} = \omega^m$ is basic, thus it is a nowhere zero multiple $\lambda.\nu$ of the transversal volume ν associated to the transversal metric (the volume ν is not necessarily holonomy invariant). This follows from the fact that these forms are both sections of the line bundle $\wedge^q Q^*$. Now consider

$$\int_M \tilde{\nu} \wedge \chi_{\mathcal{F}} = \int_M \lambda\nu \wedge \chi_{\mathcal{F}} = \int_M \lambda \cdot \mu$$

where μ is the volume form of M. Since $\int_M \mu > 0$, and λ has a fixed sign, it follows that $\int_M \tilde{\nu} \wedge \chi_{\mathcal{F}} \neq 0$. A relation $\tilde{\nu} = d\alpha$ with $\alpha \in \Omega_B^{q-1}(\mathcal{F})$ would then lead to a contradiction as in the proof of Theorem 9.21. Thus $[\tilde{\nu}] = [\omega^m] \neq 0 \in H_B^q(\mathcal{F})$.

Assume now that $[\omega^k] = 0 \in H_B^{2k}(\mathcal{F})$ for some $1 \leq k < m$. Then $\omega^k = d\alpha$ with $\alpha \in \Omega_B^{2k-1}(\mathcal{F})$ and

$$d(\omega^{m-k} \wedge \alpha) = \omega^{m-k} \wedge d\alpha = \omega^{m-k} \wedge \omega^k = \omega^m,$$

which contradicts what we just proved. ∎

Next we wish to state and prove a transversal divergence theorem, for a harmonic foliation with a holonomy invariant transversal volume ν. Let $Y \in V(\mathcal{F})$ be an infinitesimal automorphism of $\mathcal{F}$. We define then the transversal divergence $\mathrm{div}_B Y$ by

(9.24) $$\theta(Y)\nu = \mathrm{div}_B Y \cdot \nu.$$

Here we use the fact that $\theta(Y)\nu \in \Omega_B^q(\mathcal{F}) \subset \Gamma \wedge^q Q^*$. Observe that $\text{div}_B Y \in \Omega_B^o(\mathcal{F})$ and in fact depends only on $\overline{Y} = \pi(Y)$. Thus we denote it also by $\text{div}_B \overline{Y}$. We prove the following result [KTT].

9.25 TRANSVERSAL DIVERGENCE THEOREM. <u>Let $\mathcal{F}$ be a transversally oriented harmonic foliation with holonomy invariant transversal volume ν on a closed oriented Riemannian manifold</u> (M, g_M). <u>Let Y be an infinitesimal automorphism of $\mathcal{F}$ and $\overline{Y} = \pi(Y)$. Then</u>

$$\int_M \text{div}_B \overline{Y} \cdot \mu = 0.$$

PROOF. Let $\chi_{\mathcal{F}} = {}^*\nu$ be the characteristic form of $\mathcal{F}$, and assume M to be oriented by $\mu = \nu \wedge \chi_{\mathcal{F}}$. Then

$$\text{div}_B \overline{Y} \cdot \mu = (\text{div}_B \overline{Y} \cdot \nu) \wedge \chi_{\mathcal{F}} = \theta(Y)\nu \wedge \chi_{\mathcal{F}} = (d i(Y)\nu) \wedge \chi_{\mathcal{F}}$$

$$= d(i(Y)\nu \wedge \chi_{\mathcal{F}}) + (-1)^q i(Y)\nu \wedge d\chi_{\mathcal{F}}.$$

We have $\nu \in F^q$ and $i(Y)\nu \in F^{q-1}$. Since $\mathcal{F}$ is harmonic, $d\chi_{\mathcal{F}} \in F^2$ by (9.14). It follows that $i(Y)\nu \wedge d\chi_{\mathcal{F}}$ is of filtration degree $q - 1 + 2 = q + 1$, and hence vanishes. Stokes' theorem implies now the desired result. ∎

For an infinitesimal automorphism Y of a Riemannian, but not necessarily harmonic foliation the value of the integral of $\text{div}_B \overline{Y}$ is given in Theorem 9.32 below.

INFINITESIMAL AUTOMORPHISMS ONCE MORE. Let first $\mathcal{F}$ be a foliation on (M, g_M), with induced metric g_Q on Q. For an infinitesimal automorphism $Y \in V(\mathcal{F})$ we have then

$$(9.26) \quad (\theta(Y)g_Q)(Z,Z') = Yg_Q(Z,Z') - g_Q(\pi[Y,Z],Z') - g_Q(Z,\pi[Y,Z']),$$

where $Z, Z' \in \Gamma L^\perp$. Therefore

$$(\theta(Y)g_Q)(Z,Z') = Yg_M(Z,Z') - g_M([Y,Z],Z') - g_M(Z,[Y,Z'])$$

$$= g_M(\nabla_Z^M Y, Z') + g_M(Z, \nabla_{Z'}^M Y).$$

This formula generalizes (5.17).

Let now $Y = \pi^\perp(Y) + \pi(Y) = Y_L + \overline{Y}$. Then $\nabla_Z^M Y = \nabla_Z^M Y_L + \nabla_Z^M \overline{Y}$. It follows that

$$(9.28) \quad (\theta(Y)g_Q)(Z,Z') = (\theta(Y_L)g_Q)(Z,Z') + g_Q(\nabla_Z \overline{Y}, Z') + g_Q(Z, \nabla_{Z'} \overline{Y}).$$

For a Riemannian foliation the first term on the RHS vanishes, and one has the following result.

9.29. PROPOSITION. <u>Let</u> $\mathcal{F}$ <u>be a Riemannian foliation</u>. <u>For</u> $Y \in V(\mathcal{F})$ <u>and</u> $Z, Z' \in \Gamma Q$ <u>we have</u>

$$(9.30) \quad (\theta(Y)g_Q)(Z,Z') = g_Q(\nabla_Z Y, Z') + g_Q(Z, \nabla_{Z'} Y).$$

Note that the RHS involves only $\overline{Y} = \pi(Y)$.

An infinitesimal automorphism Y is <u>transversally metric</u>, if $\theta(Y)g_Q = 0$. If this holds, $Y = \pi(Y)$ is called a <u>transversal Killing field</u> (Molino [M7, 8]). For the point foliation with $L = 0$ this is the usual definition of a Killing vector field.

In terms of the basic 1-form ω given by $\omega(Z) = g_Q(Y,Z)$ for $Z \in \Gamma L^\perp$, formula (9.30) can be written in the form

(9.31) $\qquad (\theta(Y)g_Q)(Z,Z') = (\nabla_Z \omega)(Z') + (\nabla_{Z'}\omega)(Z).$

Since by definition

$$(\nabla_Z \omega)(Z') = Z\omega(Z') - \omega(\nabla_Z Z') = Zg_Q(Y,Z) - g_Q(Y,\nabla_Z Z') = g_Q(\nabla_Z Y, Z'),$$

this proves the equivalence of (9.30) and (9.31). Thus $\frac{1}{2}\theta(Y)g_Q$ is the symmetric part of the 2-linear form $\nabla \omega$ on Q.

Next we calculate the value of the integral of $\text{div}_B Y$ for $Y \in V(\mathcal{F})$ in case of a Riemannian foliation.

9.32 THEOREM. <u>Let $\mathcal{F}$ be a transversally oriented Riemannian foliation on a closed oriented Riemannian manifold (M, g_M). Let $Y \in V(\mathcal{F})$. Then</u>

$$\int_M \text{div}_B Y \cdot \mu = \int_M g_Q(\tau, Y) \cdot \mu \equiv \langle \tau, Y \rangle$$

(<u>the global scalar product of the sections τ and Y of Q</u>).

PROOF. With the notations in the proof of Theorem 9.25 we have

$$\text{div}_B Y \cdot \mu = d(i(Y)\nu \wedge \chi_{\mathcal{F}}) + (-1)^q i(Y)\nu \wedge d\chi_{\mathcal{F}}.$$

We replace $d\chi_{\mathcal{F}}$ using (9.14). Since $\varphi_0 \in F^2\Omega^{p+1}$, and $i(Y)\nu \in F^{q-1}\Omega^{p-1}$, it follows that the term $i(Y)\nu \wedge \varphi_0$ is of filtration degree $q+1$, and hence vanishes. Thus

$$\text{div}_B Y \cdot \mu = d(i(Y)\nu \wedge \chi_{\mathcal{F}}) + \kappa \wedge i(Y)\nu \wedge \chi_{\mathcal{F}}.$$

Since $\kappa \wedge \nu \in \Gamma\wedge^{q+1}Q^*$, and hence vanishes, we have further

$$0 = i(Y)\kappa \cdot \nu - \kappa \wedge i(Y)\nu.$$

Thus

$$\text{div}_B Y \cdot \mu = d(i(Y)\nu \wedge \chi_{\mathcal{F}}) + i(Y)\kappa \cdot \mu.$$

Since $\kappa(Y) = g_Q(\tau, Y)$, the desired result follows now by Stokes' Theorem. ∎

For the interaction of transversal curvature properties of a Riemannian foliation with the existence of transversal Killing fields we refer to [KT 7].

For harmonic Riemannian foliations there is a particulary intimate relationship between infinitesimal automorphisms and the transversal geometry. These are embodied in fundamental identities, involving the operators δ^*, δ occurring in the Berger-Ebin decomposition [BE], but generalized to the foliation context as in [KTT]. These identites allow formal characterizations of various geometric properties of infinitesimal automorphisms. These characterizations play a central role in the papers [KTT], [NT] and [TT].

In the case of totally geodesic foliations of codimension one Johnson and Whitt [JW 2] and Oshikiri [OS 2] show that a Killing vector field of the ambient Riemannian manifold is necessarily an infinitesimal automorphism of the foliation.

Chapter 10
FLOWS

Let $\mathcal{F}$ be a tangentially oriented foliation of dimension one on (M, g_M). Such a foliation is called a <u>flow</u>. The leaves of $\mathcal{F}$ are the integral curves of a nonsingular vector field X on M. Normalizing length shows that $\mathcal{F}$ is also given by a unit vector field T with respect to g_M. The dual 1-form $\chi \in \Omega^1(M)$ defined by

(10.1) $$\chi(Y) = g_M(T, Y) \quad \text{for} \quad Y \in \Gamma TM$$

is the characteristic form of $\mathcal{F}$. The induced metric g_L is related to χ by

(10.2) $$g_L(\lambda T, \lambda T) = \lambda^2, \quad \chi(\lambda T) = \lambda.$$

For the mean curvature vector field $\tau \in \Gamma L^\perp$, we find by (6.16)

$$\tau = \pi(\nabla_T^M T).$$

But $g_M(T, \nabla_T^M T) = \frac{1}{2} \cdot T g_M(T, T) = 0$, so that $\nabla_T^M T$ already is orthogonal to the leaves. It follows that

(10.3) $$\tau = \nabla_T^M T.$$

For the dual mean curvature 1-form we find

(10.4) $$\kappa = \theta(T)\chi.$$

PROOF. Let $Z \in \Gamma L^\perp$. Then

(10.5) $$\kappa(Z) = g_M(\nabla_T^M T, Z) = - g_M(T, \nabla_T^M Z).$$

On the other hand

$$(\theta(T)\chi)(Z) = T\chi(Z) - \chi(\theta(T)Z) = - \chi([T,Z]) = - g_M(T,[T,Z])$$

$$= g_M(T, \nabla_Z^M T - \nabla_T^M Z) = - g_M(T, \nabla_T^M Z).$$

Comparing this with (10.5) shows

$$\kappa(Z) = (\theta(T)\chi)(Z).$$

To prove (10.4) it suffices to verify that

$$(\theta(T)\chi)(T) = 0.$$

But indeed

$$(\theta(T)\chi)(T) = T\chi(T) - \chi(\theta(T)T) = 0. \blacksquare$$

From this discussion, and the results in Chapter 6, we obtain the following characterization of harmonic or geodesic flows.

10.6 THEOREM. <u>For a flow</u> $\mathcal{F}$ <u>defined by a nonsingular vector field</u> X (with normalized $T = 1/|X| \cdot X$) <u>on</u> (M, g_M), <u>the following conditions are equivalent</u>:

(i) $\mathcal{F}$ <u>is harmonic</u>;
(ii) <u>the orbits of</u> X <u>are geodesics</u>;
(iii) $\theta(T)\chi = 0$;
(iv) $\nabla_T^M T = 0$;
(v) g_L <u>is invariant under flows of vector fields orthogonal to</u> X;
(vi) $d\chi \in F^2\Omega^2(M)$;

<u>If</u> $L^\perp$ <u>is involutive, then these conditions are further equivalent to the conditions</u>:

(vii) $d\chi = 0$;
(viii) $\theta(Z)\chi = 0$ <u>for</u> $Z \in \Gamma L^\perp$.

PROOF. This follows from (6.6), (6.17), and Theorem 6.23. ∎

A flow is <u>taut</u> or <u>geodesible</u>, if there exists a Riemannian metric such that the leaves of $\mathcal{F}$ are geodesics. Note that the conditions (iii) to (viii) in Theorem 10.6 already involve a Riemannian metric. Sullivan [SU2] gave a purely topological property characteristic for such flows (see Theorem 10.17 below for a property characterizing such Riemannian flows). It is based on the following result of Gluck [GL2] and Sullivan [SU2].

10.7 THEOREM. <u>Let</u> $\mathcal{F}$ <u>be a flow given by the nonsingular vector field</u> X <u>on</u> M. <u>Then the following conditions are equivalent</u>:

(i) <u>there exists a Riemannian metric on</u> M, <u>making the orbits of</u> X <u>geodesics, and</u> X <u>of unit length</u>;

(ii) <u>there exists a 1-form</u> $\chi \in \Omega^1(M)$, <u>such that</u> $\chi(X) = 1$, <u>and</u> $\theta(X)\chi = 0$;

(iii) <u>there exists a 1-form</u> $\chi \in \Omega^1(M)$, <u>such that</u> $\chi(X) = 1$, <u>and</u> $i(X)d\chi = 0$;

(iv) <u>there exists a</u> (n-1)-<u>plane bundle</u> $E \subset TM$, <u>complementary to</u> L, <u>such that</u> $[X,Z] \in \Gamma E$ <u>for all</u> $Z \in \Gamma L$.

PROOF. (i) $\Rightarrow$ (ii): We can assume that $X = T$ with the notations at the beginning of this chapter. Define χ by (10.1). Since the orbits of T are geodesics, we have $\nabla^M_T T = 0$. Thus by Theorem 10.6, part (iii), we have $\theta(T)\chi = 0$.

(ii) $\Rightarrow$ (iii): We have

$$0 = \theta(X)\chi = i(X)d\chi + di(X)\chi = i(X)d\chi.$$

(iii) $\Rightarrow$ (iv): Let $E = \ker \chi$, and $Z \in \Gamma E$. Then

$$\chi[X,Z] = -d\chi(X,Z) + X\chi(Z) - Z\chi(X) = 0,$$

and $[X,Z] \in \Gamma E$.

(iv) $\Rightarrow$ (i): Let g_M be a metric, such that $X = T$ is of unit length, and E the orthogonal complement of T, with an arbitrary choice of metric on E. It suffices to show that $\kappa(Z) = 0$ for $Z \in \Gamma E$. But

$$\kappa(Z) = g_M(\nabla^M_T T, Z) = -g_M(T, \nabla^M_T Z) = -g_M(T, \nabla^M_Z T + [T,Z]).$$

Since by assumption $[T,Z] \in \Gamma E$, it follows that

$$\kappa(Z) = -g_M(T, \nabla^M_Z T) = -\frac{1}{2} Z g_M(T,T) = 0. \quad \blacksquare$$

10.8 COROLLARY [SU 2]. Let X be a nonsingular vector field on M. Then the following conditions are equivalent:
(i) the flow of X is geodesible;
(ii) there exists a 1-form $\chi \in \Omega^1(M)$ such that $\chi(X) > 0$ and $i(X)d\chi = 0$.

PROOF. (i) $\Rightarrow$ (ii): Let g_M be a metric such that the orbits of X are geodesics. Normalize X to a unit vector field. Then (i) $\Rightarrow$ (iii) in Theorem 10.7 implies the desired result. (ii) $\Rightarrow$ (i): Let χ be given as indicated. Then $X' = \frac{1}{\chi(X)} \cdot X$ satisfies property (iii) in Theorem 10.7. The implication (iii) $\Rightarrow$ (i) in Theorem 10.7 implies that the orbits of X' are geodesics (and X' is of unit length). But the orbits of X and X' are the same. ∎

Let X be a nonsingular Killing vector field on (M, g_M). If we renormalize the metric by

(10.9)
$$\bar{g}_M | L = \frac{1}{|X|^2} \cdot g_M | L$$

$$\bar{g}_M | L^\perp = g_M | L^\perp$$

then X is still a Killing vector field for $(M, \bar{g}_M)$, and moreover of unit length in the new metric.

A nonsingular Killing vector field clearly defines a Riemannian flow.

A flow $\mathcal{F}$ on M is <u>isometric</u>, if there exists a Riemannian metric g_M such that $\mathcal{F}$ is given by the orbits of a nonsingular Killing vector field with respect to g_M. Such a flow is necessarily Riemannian. We state the following fact (Carrière [CA 1,2]).

10.10 PROPOSITION. Let $\mathcal{F}$ be a Riemannian flow on M. Then the following conditions are equivalent:

(i) $\mathcal{F}$ is isometric;

(ii) $\mathcal{F}$ is geodesible.

PROOF. (i) $\Rightarrow$ (ii): Let X be a Killing vector field of g_M, such that $\mathcal{F}$ is given by the orbits of X. Then the orthogonal complement E of X is preserved by the flow of X, and hence property (iv) of Theorem 10.7 is satisfied. (Renormalizing the metric by (10.9), X is a unit vector field, and the orbits are geodesics by the proof of (iv) $\Rightarrow$ (i) in Theorem 10.7).

(ii) $\Rightarrow$ (i): Let $\mathcal{F}$ be a geodesible Riemannian flow. There exists a bundle-like metric and a unit vector field X, such that properties (i) and (iv) with $E = L^\perp$ of Theorem 10.7 hold. For such a metric X is a Killing vector field. ∎

For the proof of the implication (ii) $\Rightarrow$ (i), we used the assumption that the given flow is Riemannian.

For these special Riemannian flows, the basic cohomology can be described in more detail [KT 11]. Let X be a nonsingular Killing vector field on a closed Riemannian manifold M. By renormalizing the metric as in (10.9), if necessary, we can assume that $X = T$ is a unit vector field. The closure of exp tX in the compact isometry group of g_M is compact and abelian, hence a torus G. For the basic forms of the foliation $\mathcal{F}$ by the orbits of T we have then

$$\Omega_B(\mathcal{F}) \subset \Omega(M)^G,$$

where $\Omega(M)^G$ denotes the G-invariant forms on M. On the other hand the operator $i(T)$ on $\Omega(M)$ gives rise to a map of degree -1

(10.11) $$\Omega^{\cdot}(M)^G \xrightarrow{i(T)} \Omega_B^{\cdot-1}(\mathcal{F}).$$

Namely for $\omega' = i(T)\omega$, $\omega \in \Omega(M)^G$ we have

$$i(T)\omega' = i(T)^2\omega = 0, \quad \theta(T)\omega' = \theta(T)i(T)\omega = i(T)\theta(T)\omega = 0.$$

This map is compatible with differentials (up to sign), since $di(T) + i(T)d = \theta(T)$, which vanishes on $\Omega(M)^G$. We further show that this map is surjective. Let χ be the characteristic form. Consider for $\omega \in \Omega_B^{r-1}(\mathcal{F})$ the r-form $\chi \wedge \omega$. Then

$$\theta(T)(\chi \wedge \omega) = \theta(T)\chi \wedge \omega \pm \chi \wedge \theta(T)\omega = 0,$$

and $\chi \wedge \omega \in \Omega(M)^G$. Moreover

$$i(T)(\chi \wedge \omega) = i(T)\chi \cdot \omega \pm \chi \wedge i(T)\omega = \omega,$$

which proves the surjectivity of the map (10.11). As a consequence one has the following result [GHV].

10.12 PROPOSITION. <u>There is an exact sequence of complexes</u>

$$0 \longrightarrow \Omega_B^{\cdot}(\mathcal{F}) \hookrightarrow \Omega^{\cdot}(M)^G \xrightarrow{i(T)} \Omega_B^{\cdot-1}(\mathcal{F}) \longrightarrow 0.$$

PROOF. It only remains to verify the exactness in the middle. For a basic form $\omega, i(T)\omega = 0$, which proves im $\subset$ ker $i(T)$. Let conversely $\omega \in \Omega(M)^G$ with $i(T)\omega = 0$. Since $\theta(T)\omega = 0$, $\omega \in \Omega_B(\mathcal{F})$. ∎

Since $H(\Omega^\cdot(M)^G) \xrightarrow{\sim} H(\Omega^\cdot(M))$, as for any compact Lie group acting on M, the following facts are a consequence of Proposition 10.12.

10.13 THEOREM. Let T be a nonsingular Killing vector field on the closed manifold (M^n, g_M).
(i) There is a long exact cohomology sequence

$$\cdots \longrightarrow H_B^r(\mathcal{F}) \longrightarrow H_{DR}^r(M) \xrightarrow{i(T)_*} H_B^{r-1}(\mathcal{F}) \xrightarrow{\Delta} H_B^{r+1}(\mathcal{F}) \longrightarrow \cdots$$

with connecting homomorphism Δ.
(ii) The groups $H_B^r(\mathcal{F})$ are all finite-dimensional for $0 \leq r \leq n - 1$ (and 0 otherwise).
(iii) The Euler characteristic

$$\chi_B(\mathcal{F}) = \sum_{r=0}^{n-1} (-1)^r \dim H_B^r(\mathcal{F})$$

is a well-defined integer.

As an illustration consider the case when M^n has the DeRham cohomology of a sphere S^n, $n = 2k + 1$. If T is a nonsingular Killing vector field, then one concludes recursively that $H_B^r(\mathcal{F}) \cong H_{DR}^r(\mathbb{P}^k\mathbb{C})$ for $r = 0, \ldots, 2k$.

We note that the cohomology sequence in (10.13) ends with the exact sequence

(10.14) $$0 \longrightarrow H_{DR}^n(M) \xrightarrow[\cong]{i(T)_*} H_B^{n-1}(\mathcal{F}) \longrightarrow 0.$$

Thus for a closed oriented manifold, the volume form μ of M maps to a generator $i(T)\mu$ of $H_B^{n-1}(\mathcal{F}) \cong \mathbb{R}$. This is a special case of the situation described earlier in Theorem 9.21 and Corollary 9.22.

The boundary map Δ in 10.13 is given as follows. Let $\omega \in \Omega_B^{r-1}(\mathcal{F})$ be closed. Then $i(T)(\chi \wedge \omega) = \omega$, as proved before. Thus $\Delta[\omega]$ is represented by $d(\chi \wedge \omega) = d\chi \wedge \omega \in \Omega_B^{r+1}(\mathcal{F})$, i.e.

(10.15) $$\Delta[\omega] = [d\chi] \cdot [\omega], \text{ where } [d\chi] \in H_B^2(\mathcal{F}).$$

We can define a DG-algebra $E = \wedge(\chi) \otimes \Omega_B$ with $d_E(\chi \otimes 1) = 1 \otimes d\chi$, $d_E(1 \otimes \omega) = 1 \otimes d_B\omega$. The canonical map $E \hookrightarrow \Omega(M)$ induces an isomorphism on the E_3-level, and thus an isomorphism [KT 11]

(10.16) $$H^{\cdot}(E) \cong H_{DR}^{\cdot}(M).$$

In the presence of an involutive complement to the flow, it follows by part (vii) of Theorem 10.6 that $d\chi = 0$, and the coboundary map (10.15) is trivial.

As pointed out in [SA], the vanishing of $[d\chi] \in H_B^2(\mathcal{F})$ conversely implies that the isometric flow $\mathcal{F}$ __is__ transverse to a fibration $M \longrightarrow S^1$. To see this, let $\alpha \in \Omega_B^1(\mathcal{F})$ such that $d\chi = d_B\alpha$. Then $\omega = \chi - \alpha \in \Omega^1(M)$ satisfies $d\omega = 0$ and

$$\omega(T) = \chi(T) - \alpha(T) = 1.$$

Thus ω is nonsingular, and by Tischler's Theorem 4.6 there is a fibration $f : M \longrightarrow S^1$ with $\omega' = f^*d\theta$ arbitrarily close to ω. Thus $\omega'(T) > 0$ and $\mathcal{F}$ is transverse to $f : M \longrightarrow S^1$.

From (10.14) we know that for an isometric (equivalently, geodesible) flow $\mathcal{F}$ on a closed oriented manifold M^n we have $H_B^{n-1}(\mathcal{F}) \cong \mathbb{R}$. Thus the following result of Molino and Sergiescu [MS] is very appealing.

10.17 THEOREM. Let $\mathcal{F}$ be a Riemannian flow on a closed oriented manifold M^n. Then the following conditions are equivalent:
(i) $\mathcal{F}$ is isometric;
(ii) $H_B^{n-1}(\mathcal{F}) \neq 0$.

In view of the remark above, these conditions are further equivalent to:
(ii') $H_B^{n-1}(\mathcal{F}) \cong \mathbb{R}$.

The interest of this characterization of isometric flows resides in the fact, that it involves no metric data whatsoever.

For simply-connected (and closed oriented) M, these conditions are always satisfied (see [GH 2, Thm. B], and [KT 11, Cor. 4.24] for the case of a bundle-like metric with basic mean curvature form).

It is of interest to observe that Carrière [CA 1,2] has given examples of Riemannian flows on oriented closed 3-manifolds with $H_B^2(\mathcal{F}) = 0$. This led him to conjecture Theorem 10.17. Carrière's examples are the torus fibrations T_A over the circle discussed in Chapter 4. A typical example is given by $A = \begin{bmatrix} 1 & 1 \\ 1 & 2 \end{bmatrix} \in SL(2,\mathbb{Z})$. The eigenspace corresponding to the eigenvalue $1/2 \cdot (3 - \sqrt{5})$ gives rise to a Riemannian flow on $T_A = \mathbb{R} \times_A T^2$. The basic cohomology of this flow can be calculated, because the transversal structure is modeled on the affine group GA associated to A, and acting on $\mathbb{R}^2$. The relevant fact is then that $H_B(\mathcal{F})$ is isomorphic to the cohomology of the (left-) invariant forms on GA invariant under a compact subgroup K, the closure of the holonomy group in GA (see also Blumenthal [BL 2]). Then Carrière shows explicitly, that a K-invariant 2-form on GA is the boundary of a K-invariant 1-form, which proves $H_B^2(\mathcal{F}) = 0$.

In the attempt to classify Riemannian flows (which succeeds on manifolds of low dimension), the following result [CA 1,2] is a key result.

10.18 THEOREM. <u>Let $\mathcal{F}$ be a Riemannian flow with dense orbits on a closed manifold M^n. Then M^n is diffeomorphic to the torus T^n, and the flow $\mathcal{F}$ is (smoothly) conjugate to a linear flow on T^n.</u>

The last statement means, that there exists a smooth diffeomorphism $M^n \longrightarrow T^n$ taking the leaves of $\mathcal{F}$ to the orbits of a linear flow on T^n.

This theorem leads to the classification of Riemannian flows on closed 3-manifolds [CA 1,2]. For the classification of Riemannian flows on closed 4-manifolds, see [AM].

CHAPTER 11

LIE FOLIATIONS

The proofs of Theorems 10.17 and 10.18, as well as many other results on Riemannian foliations, rely substantially on the structure theory for Riemannian foliations developed by Molino [M 8] in arbitrary codimension. It is based on several fundamental observations. The first is that the canonical lift $\hat{\mathcal{F}}$ of a Riemannian foliation $\mathcal{F}$ to the bundle P of orthonormal frames of Q, is a transversally parallelizable Riemannian foliation. The canonical lift $\hat{\mathcal{F}}$ on P is a foliation of the same dimension as $\mathcal{F}$ on M, and invariant under the action of the orthogonal structural group of P. Now let M be closed and oriented, and consider on P the closures of the leaves of $\hat{\mathcal{F}}$. The second fundamental observation is that these closures form the fibers of a fibration $X_0 \longrightarrow P \overset{\pi}{\longrightarrow} W$, over the space W of orbit closures, with typical fiber X_0. The foliation $\hat{\mathcal{F}}$ induces on X_0, and on each fiber of π, a foliation with dense leaves, that is transversally modeled on a Lie group G with translations as transition functions. These are the Lie foliations previously studied by Fedida [F] and Molino [M 5,6]. The Lie algebra of this group G is another structural invariant of the foliation. With the help of this structure theorem, many questions on Riemannian foliations can be reduced to questions on Lie foliations, by passing to the bundle of transversal orthonormal frames.

Molino has recently generalized this structure theorem to certain types of singular Riemannian foliations [M 11].

The precise definition of a Lie foliation is as follows [F] [M 5,6]. Let $\mathfrak{g}$ be a real Lie algebra, and $\omega \in \Omega^1(M,\mathfrak{g})$ a $\mathfrak{g}$-valued 1-form. ω is a Maurer-Cartan form (MC-form), if it satisfies the MC-equation $d\omega + \frac{1}{2}[\omega,\omega] = 0$. In other words the (formal) curvature vanishes. If $\omega_x : T_xM \longrightarrow \mathfrak{g}$ is surjective for all $x \in M$, then $L_x = \ker \omega_x$ defines a foliation of codimension $q = \dim \mathbf{g}$ on M. Such a foliation is called a $\mathbf{g}$-Lie foliation. For $\mathbf{g} = \mathbb{R}$ (or more generally for abelian $\mathbf{g}$), the MC-equation reduces to $d\omega = 0$.

The involutivity of L follows immediately, namely $i(X)\omega = 0$, $i(Y)\omega = 0$ implies

$$d\omega(X,Y) = -\omega[X,Y],$$

which vanishes by the MC-equation.

A flat connection on a principal G-bundle $P \longrightarrow M$ is given by a MC-form on P, with values in the Lie algebra $\mathbf{g}$ of G. Conversely consider a MC-form ω on M with values in $\mathbf{g}$, and G a Lie group with Lie algebra $\mathbf{g}$. Then a flat connection on the principal G-bundle $M \times G$ is defined by the formula

$$\tilde{\omega}_{(x,g)}(X_x,Y_g) = \mathrm{Ad}(g^{-1})\omega(X_x) + Y,$$

where $X \in \Gamma TM$, $Y \in \mathbf{g}$, and Y_g denotes the value at g of the corresponding left-invariant vectorfield on G. The trivial section s of $M \longrightarrow G$ pulls this form $\tilde{\omega}$ back to ω. This establishes a bijective correspondence between MC-forms on M and flat connections on $M \times G$. The Lie foliations on M are distinguished among these by the nonsingularity of $L = \ker \omega \subset TM$. This means that $\dim L = $ constant, and the codimension q equals $\dim \mathbf{g}$.

To show that a $\mathbf{g}$-Lie foliation is necessarily Riemannian, consider a basis $e_1,\ldots,e_q$ of $\mathbf{g}$. Let $s_1,\ldots,s_q \in \Gamma\mathbb{Q}$ such that for each $x \in M$ we have $\omega(s_\alpha)_x = e_\alpha$. Since $\omega_x : T_xM \xrightarrow{\sim} \mathbf{g}$, it follows that $s_1,\ldots,s_q$ are a framing of $\mathbb{Q}$. Let Y_α be a lift of s_α to TM. Thus $Y_\alpha = s_\alpha$ with an earlier notation. We will see in Lemma 11.1 that $Y_\alpha \in V(\mathcal{F})$, i.e. $[X, Y_\alpha] \in \Gamma L$ for $X \in \Gamma L$. The Euclidean metric on $\mathfrak{g}$, with $e_1, \ldots, e_q$ as orthonormal basis, defines a Riemannian metric g_Q on Q. We have then for $X \in \Gamma L$

$$(\theta(X)g_Q)(s_\alpha, s_\beta) = Xg_Q(s_\alpha, s_\beta) - g_Q(\pi[X, Y_\alpha], s_\beta) - g_Q(s_\alpha, \pi[X, Y_\beta])$$

$$= Xg_Q(s_\alpha, s_\beta) \ .$$

Since $g_Q(s_\alpha, s_\beta) = \delta_{\alpha\beta}$, this expression vanishes, and $\mathcal{F}$ is a Riemannian foliation.

We have used the following fact.

11.1 LEMMA. <u>With the notations above we have</u>
(i) $Y_\alpha \in V(\mathcal{F})$, $\alpha = 1, \ldots, q$;
(ii) $\omega[Y_\alpha, Y_\beta] = [e_\alpha, e_\beta]$.

PROOF. (i) Let $X \in \Gamma L$. We have to show $\omega[Y_\alpha, X] = 0$. But

$$d\omega(Y_\alpha, X) = Y_\alpha \omega(X) - X\omega(Y_\alpha) - \omega[Y_\alpha, X] = - \omega[Y_\alpha, X] ,$$

since clearly $\omega(X) = 0$, and $\omega(Y_\alpha) = \omega(Y_\alpha) = $ constant, and thus also $X\omega(Y_\alpha) = 0$. By the MC-equation

$$d\omega(Y_\alpha, X) = -\frac{1}{2}[\omega, \omega](Y_\alpha, X) = - [\omega(Y_\alpha), \omega(X)] = 0,$$

and thus $\omega[Y_\alpha, X] = 0$.

(ii) By a similar reasoning

$$\omega[Y_\alpha, Y_\beta] = \omega[Y_\alpha, Y_\beta] = -d\omega(Y_\alpha, Y_\beta) = \frac{1}{2}[\omega,\omega](Y_\alpha, Y_\beta) = [\omega(Y_\alpha), \omega(Y_\beta)] = [e_\alpha, e_\beta]. \blacksquare$$

Thus we can identify g with ΓQ^L and s_α wtih e_α.

The developing map of a g-Lie foliation on M is described as follows. Let $\tilde{M} \xrightarrow{p} M$ be the universal covering of M, and G the simply connected Lie group with Lie algebra g. Let $\tilde{x}_0$ be a basepoint in $\tilde{M}$, and $x_0 = p(\tilde{x}_0)$. For any path γ from $\tilde{x}_0$ to a point $\tilde{x}$ in $\tilde{M}$, the parallel transport τ_γ in the flat principal G-bundle $\tilde{M} \times G \longrightarrow \tilde{M}$ is well-defined. It maps $(\tilde{x}_0, g)$ to the point $(\tilde{x}, g)$, and by definition

$$f_\omega : \tilde{M} \longrightarrow G, \ f_\omega(\tilde{x}) = g.$$

This value depends only on $\tilde{x}$ (and not on the choice of γ), because another path γ' from $\tilde{x}_0$ to $\tilde{x}$ is homotopic to γ, thus $\tau_{\gamma'} = \tau_\gamma$ as the curvature is zero. The surjectivity of ω turns f_ω into a submersion. If M is closed, this map can be shown to be a locally trivial fibration. The crucial point is the existence of a Riemannian metric, turning f_ω into a Riemannian submersion. Such a metric is constructed by defining first a holonomy invariant Riemannian metric g_Q on Q as before, and setting $g_M = g_L \oplus g_Q$ with g_L an arbitrary bundle metric on L.

The holonomy homomorphism

$$h_\omega : \pi_1(M, x_0) \longrightarrow G$$

is similarly defined in the flat principal bundle $M \times G \longrightarrow M$, and for loops in M based at x_0. The developing map f_ω is equivariant with respect to h, i.e.

$$f_\omega(\tilde{x}\gamma) = f_\omega(\tilde{x}) \cdot h_\omega(\gamma).$$

It follows that f_ω induces a map $M \longrightarrow G/\operatorname{im} h_\omega$.

Tischler's Theorem 4.6 corresponds to the case $g = \mathbb{R}$ and $d\omega = 0$. In that case h_ω is the period map of ω, and $\operatorname{im} h_\omega \subset \mathbb{R}$ the group of periods of ω.

Returning to a general g-Lie foliation on M with developing map $\tilde{M} \longrightarrow G$, it follows that it can be defined by an atlas $\mathcal{U} = \{U_\alpha\}$ of distinguished charts, submersions $f_\alpha : U_\alpha \longrightarrow G$, and transition functions which are left translations in G. Thus we have a G-foliation, where G is identified with its group of left translations, which in turn is a subgroup of the affine automorphism group of the left parallelism of G. To see that $G \subset GL(q)$, $q = \dim G$, observe that an affine automorphism is completely determined by its value and the derivative at one point.

If the holonomy homomorphism h_ω has its image in H, where H is a closed subgroup of G, then the developing map induces a submersion $M \longrightarrow G/H$. Note that the transition functions in this case are given by the action of G on the homogeneous space G/H.

This is an example of a transversally homogeneous foliation, as discussed in [BL 1]. Such a foliation is given by local submersions to G/H, H a closed connected subgroup of G, with transition functions given by the action of G on G/H. As shown in [BL 1], for closed M the developing map is then a submersion $\tilde{M} \longrightarrow G/H$.

Consider again a g-Lie foliation $\mathcal{F}$, given by a Maurer-Cartan form in $\Omega^1(M,g)$, and thus $\Gamma Q^L \cong g$. If the leaves of $\mathcal{F}$ are dense, there is a homomorphism

$$\Omega_B^{\cdot}(\mathcal{F}) \longrightarrow \wedge^{\cdot} \mathfrak{g}^*$$

into the Chevally-Eilenberg complex of $\mathfrak{g}$. This homomorphism is defined as follows. A basic form on M lifts to a form on the universal covering $\tilde{M}$, basic with respect to the lifted foliation, i.e. basic with respect to the developing map $\tilde{M} \longrightarrow G$, a submersion. The resulting form on G is invariant under the action by left translations of the holonomy group H, the image of the holonomy homomorphism. The density of the leaves implies that $\bar{H} = G$. Thus the resulting form on G is left invariant, hence an element in $\wedge^{\cdot} \mathfrak{g}^*$. This construction gives rise to an isomorphism

$$H_B^{\cdot}(\mathcal{F}) \xrightarrow{\sim} H^{\cdot}(\mathfrak{g}, \mathbb{R})$$

with the Lie algebra cohomology of $\mathfrak{g}$.

CHAPTER 12

TWISTED DUALITY

Throughout this chapter $\mathcal{F}$ denotes a transversally oriented Riemannian foliation on a closed oriented manifold M. We discuss a duality theorem for the cohomology of basic forms [KT 11, 13].

Let g_M be a bundle-like metric inducing g_Q on $L^\perp \cong Q$. Besides the *-operator on $\Omega^{\cdot}(M)$ associated to g_M, there is also a star operator

(12.1) $$ \bar{*} : \Omega^r_B(\mathcal{F}) \longrightarrow \Omega^{q-r}_B(\mathcal{F}) .$$

Since $\mathcal{F}$ is locally given by Riemannian submersions $f : U \longrightarrow N$ into an oriented manifold N, with isometries of N as local transition functions, it follows that the star operator on the q-dimensional manifold N transports to the $\mathcal{F}$-basic forms. The relationships between $*$ in $\Omega^{\cdot}(M)$ and $\bar{*}$ in $\Omega^{\cdot}_B(\mathcal{F})$ are described by the following formulas:

(12.2) $$ *\alpha = (-1)^{p(q-r)} \bar{*}(\alpha \wedge \chi_{\mathcal{F}}) $$

(12.3) $$ *\alpha = \bar{*}\alpha \wedge \chi_{\mathcal{F}} $$

for $\alpha \in \Omega^r_B(\mathcal{F})$, and $\chi_{\mathcal{F}}$ the characteristic form of $\mathcal{F}$. For the transversal volume form $\nu \in \Omega^q_B(\mathcal{F})$, the last formula yields $*\nu = \chi_{\mathcal{F}}$ (see the orientation conventions at the end of Chapter 6).

In $\Omega^r_B(\mathcal{F})$ we have the natural scalar product

(12.4) $$ <\alpha,\beta>_B = \int_M \alpha \wedge \bar{*}\beta \wedge \chi_{\mathcal{F}} . $$

In view of (12.3), this is the restriction of the usual scalar product on $\Omega^r(M)$ to the subspace $\Omega_B^r(\mathcal{F})$.

The subsequent calculations in this chapter are done for a Riemannian foliation, with a bundle-like metric g_M with mean curvature form $\kappa \in \Omega_B^1(\mathcal{F})$. For $\kappa \in \Omega_B^1(\mathcal{F})$ we have the following crucial property [KT 11]:

(12.5) $$d\kappa = 0.$$

PROOF. Since $d\kappa \in \Omega_B^2$, we have $d\kappa = {}^*\alpha$ for some $\alpha \in \Omega_B^{q-2}$. It suffices to show $\alpha = 0$. But for the square of the global norm we have

$$\|\alpha\|^2 = \int_M \alpha \wedge {}^*\alpha \wedge \chi_\mathcal{F} = \int_M \alpha \wedge d\kappa \wedge \chi_\mathcal{F}.$$

By (9.14) we have $d\chi_\mathcal{F} + \kappa \wedge \chi_\mathcal{F} = \varphi_0 \in F^2 \Omega^{p+1}$, and thus by differentiation

$$d\kappa \wedge \chi_\mathcal{F} - \kappa \wedge d\chi_\mathcal{F} = d\varphi_0.$$

Using (9.14) again this implies

$$d\kappa \wedge \chi_\mathcal{F} = \kappa \wedge (\varphi_0 - \kappa \wedge \chi_\mathcal{F}) + d\varphi_0 = \kappa \wedge \varphi_0 + d\varphi_0.$$

On the RHS the first term is a form $\psi = \kappa \wedge \varphi_0 \in F^3$, while $d\varphi_0 \in F^2$. It follows that

$$\|\alpha\|^2 = \int_M \alpha \wedge \psi + \int_M \alpha \wedge d\varphi_0.$$

On the RHS the first integrand is of filtration degree $(q - 2) + 3 = q + 1$, hence vanishes. The second integrand is up to sign of the form

$$d(\alpha \wedge \varphi_0) - d\alpha \wedge \varphi_0.$$

The second term is of filtration degree $(q-1) + 2 = q+1$, and hence trivial, while the other term vanishes by integration. Thus $\|\alpha\|^2 = 0$ and hence $\alpha = 0$. ∎

The cohomology class $[\kappa] \in H_B^1(\mathcal{F})$ so obtained is of great interest [KT 11].

12.6 PROPOSITION. <u>Let</u> $[\kappa] = 0$. <u>Then</u> $\mathcal{F}$ <u>is taut, i.e. the bundle-like metric</u> g_M <u>can be modified to a bundle-like metric</u> g_M' <u>with minimal leaves.</u>

PROOF. Let $\kappa = df$ with $f \in \Omega_B^0(\mathcal{F})$. This can be written $\kappa = d \log \lambda$, for $\lambda = e^f \in \Omega_B^0(\mathcal{F})$. Define

(12.7) $$g_M' = \lambda^{2/p} \cdot g_L \oplus g_Q.$$

Then the mean curvature form κ' associated to g_M' is calculated to be

(12.8) $$\kappa' = \kappa - d \log \lambda = 0. \quad \blacksquare$$

By Proposition 9.9 we have an injective map $H_B^1(\mathcal{F}) \longrightarrow H_{DR}^1(M)$. This yields the following result ([GH 2], [KT 11] for the case of a bundle-like metric with $\kappa \in \Omega_B^1(\mathcal{F})$).

12.9 COROLLARY. <u>Let</u> M <u>be closed and simply connected. Then every Riemannian foliation on</u> M <u>is taut.</u>

Next we calculate the formal adjoint

$\delta_B : \Omega_B^r(\mathcal{F}) \longrightarrow \Omega_B^{r-1}(\mathcal{F})$ of $d_B = d : \Omega_B^{r-1}(\mathcal{F}) \longrightarrow \Omega_B^r(\mathcal{F})$

with respect to the scalar product (12.4). It is defined by

$$\langle d_B \alpha, \beta \rangle_B = \langle \alpha, \delta_B \beta \rangle_B.$$

The mean curvature form appears in an essential role [KT 9,11].

12.10 THEOREM. <u>The formal adjoint of</u> d_B <u>in</u> $\Omega_B^{\cdot}(\mathcal{F})$ <u>with respect to</u> $\langle , \rangle_B$ <u>is the operator</u>

(12.11) $\qquad \delta_B = (d_B - \kappa \wedge)^* : \Omega_B^r(\mathcal{F}) \longrightarrow \Omega_B^{r-1}(\mathcal{F}),$

<u>where for</u> $\beta \in \Omega_B^r(\mathcal{F})$

(12.12) $\qquad (d_B - \kappa \wedge)^* \beta \equiv (-1)^{q(r+1)+1} * (d_B - \kappa \wedge) * \beta.$

PROOF. Let $\alpha \in \Omega_B^{r-1}$, $\beta \in \Omega_B^r$. Then

$$\langle d_B \alpha, \beta \rangle_B = \int_M d\alpha \wedge (*\beta \wedge \chi_{\mathcal{F}}) = (-1)^r \int_M \alpha \wedge d(*\beta \wedge \chi_{\mathcal{F}})$$

$$= (-1)^r \int_M \alpha \wedge (d*\beta \wedge \chi_{\mathcal{F}} + (-1)^{q-r} *\beta \wedge d\chi_{\mathcal{F}}).$$

By (9.14) we can replace $d\chi_{\mathcal{F}}$ by $-\kappa \wedge \chi_{\mathcal{F}}$, up to a form $\varphi_0 \in F^2 \Omega^{p+1}$. Since $\alpha \in F^{r-1}$ and $*\beta \in F^{q-r}$, it follows that

$$\alpha \wedge *\beta \wedge d\chi_{\mathcal{F}} = \alpha \wedge *\beta \wedge (-\kappa \wedge \chi_{\mathcal{F}}),$$

because the difference is a form of filtration degree $(r-1) + (q-r) + 2 = q + 1$, and hence vanishes. Thus

$$\langle d_B\alpha, \beta\rangle_B = (-1)^r \int_M \alpha \wedge [d_B^*\beta - (-1)^{q-r} *\beta \wedge \kappa] \wedge \chi_{\mathcal{F}}$$

$$= (-1)^r(-1)^{(q-r+1)(r-1)} \int_M \alpha \wedge *[*d_B^*\beta - *(\kappa \wedge *\beta)] \wedge \chi_{\mathcal{F}}$$

$$= (-1)^{q(r+1)+1} \int_M \alpha \wedge *[*(d_B - \kappa\wedge)*\beta] \wedge \chi_{\mathcal{F}}$$

$$= \int_M \alpha \wedge *(d_B - \kappa\wedge)^*\beta \wedge \chi_{\mathcal{F}} \;,$$

which establishes the desired result. ∎

12.13 COROLLARY. <u>For the transversal invariant volume ν, we obtain in particular</u>

(12.14) $$\delta_B \nu = *\kappa.$$

We introduce the twisted differential [KT 11, 13]

(12.15) $$d_\kappa = d_B - \kappa\wedge.$$

Formula (12.11) reads then $\delta_B = (d_\kappa)^*$. Since $d\kappa = 0$, $(d_\kappa)^2 = 0$ and hence $\delta_B^2 = 0$. Note that δ_B is obtained by modifying the codifferential associated to the transversal Riemannian metric by the operator $(-\kappa\wedge)^*$ of order zero (and degree -1). In terms of the operator d_κ we have

(12.16) $\langle d_B \alpha, \beta \rangle_B = \langle \alpha, (d_\kappa)^* \beta \rangle_B$ for $\alpha \in \Omega_B^{r-1}(\mathcal{F})$, $\beta \in \Omega_B^r(\mathcal{F})$.

We define further

(12.17) $d_B^* \beta = (-1)^{q(r+1)+1} * d_B * \beta$ for $\beta \in \Omega_B^r(\mathcal{F})$.

Then by a calculation like the one establishing (12.11) we find

(12.18) $\langle d_\kappa \alpha, \beta \rangle_B = \langle \alpha, d_B^* \beta \rangle_B$ for $\alpha \in \Omega_B^{r-1}(\mathcal{F})$, $\beta \in \Omega_B^r(\mathcal{F})$.

Formulas (12.16) and (12.18) express the facts that d_B, $\delta_B = (d_\kappa)^*$ and d_κ, $\delta_\kappa = d_B^*$ are two pairs of mutually adjoint operators with respect to $\langle\ ,\ \rangle_B$. We have therefore two Laplacians

(12.19) $\Delta_B = d_B \delta_B + \delta_B d_B$

(12.20) $\Delta_\kappa = d_\kappa \delta_\kappa + \delta_\kappa d_\kappa$.

They are related by

(12.21) $*\Delta_B = \Delta_\kappa *$,

and thus it suffices to consider Δ_B. The harmonic basic r-forms $\mathcal{H}_B^r$ are those satisfying $\Delta_B \omega = 0$. The following generalization of the usual De Rham-Hodge decomposition holds.

12.22 THEOREM. Let $\mathcal{F}$ be a transversally oriented Riemannian foliation on a closed oriented manifold (M, g_M). Assume g_M to be bundle-like with $\kappa \in \Omega_B^1(\mathcal{F})$. Then there is a decomposition into mutually orthogonal subspaces

$$\Omega_B^r \cong \operatorname{im} d_B \oplus \operatorname{im} \delta_B \oplus \mathcal{H}_B^r$$

with finite-dimensional $\mathcal{H}_B^r$.

Proofs have appeared in [EH 2] and [KT 16]. The finite-dimensionality of $\mathcal{H}_B^r$ was established in [ESH]. The proof in [EH 2] is based on Molino's structure theorem for Riemannian foliations [M 8] as sketched on page 143. It tracks the De Rham-Hodge decomposition for the basis of the associated adherence foliation on the normal frame bundle through the corresponding spectral sequences.

The idea of the proof in [KT 16], is to construct a strongly elliptic operator on all forms, which on basic forms restricts to the basic Laplacian Δ_B. The ordinary Laplacian Δ does not have this property (except for particularly simple foliations). The proper extension is an operator $\Delta - \tilde{\eta}$, where Δ is the ordinary Laplacian, and $\tilde{\eta}$ an explicitly defined operator of order (not exceeding) one, and preserving the degree of forms. This extension is not necessarily self-adjoint.

Before continuing with the description of this proof, we wish to point out that in the application discussed in Chapter 13 this approach to the De Rham-Hodge theory proved successful, while the use of the existence of a decomposition alone was not conclusive. It proved necessary to return to the fundamental estimates, involving the expressions resulting from the above explicit construction.

It should further be pointed out that the De Rham-Hodge decomposition depends on the particular metric involved, which as stated above is assumed to have a basic mean curvature form. Note that the cohomology spaces $H_B(\mathcal{F})$ do not involve any metric data, while the Theorem implies an isomorphism $H_B^r(\mathcal{F}) \cong \mathcal{H}_B^r$ (see Theorem 12.30 below), for the given particular metric.

Returning to the proof outline, the next point is that the known coercivity of the bilinear form associated to the strongly elliptic $\Delta - \tilde{\eta}$ implies the corresponding property for the operator Δ_B. By the abstract theorem of [E], this leads to the existence of weak solutions for the usual Poisson equations. A technical difficulty encountered at this stage is the verification of the Rellich and Sobolev property for the Sobolev chain $H_s(\Omega_B)$, $s \geq 0$ of the basic complex. The remaining part of the proof consists in establishing a regularity theorem, which leads to the actual solvability of the relevant Poisson equations. Note that one cannot simply apply the usual arguments directly, because the basic forms do not constitute all sections of a vector bundle, but rather the intersection of the kernels of Lie derivative operators within all sections of such a bundle.

Next we describe the extension of Δ_B in detail. For this purpose it is useful to introduce a bounded linear operator $\tilde{\gamma} : \Omega^r \longrightarrow \Omega^{p+q-r+1}$ of order 0 defined by

$$(12.23) \qquad \tilde{\gamma}(\omega) = (-1)^{(q+1)(p+r)+1} * (\omega \wedge \chi_{\mathcal{F}}) \wedge \varphi_0$$

where φ_0 is given as in (9.14). This formula restricts by (12.2) on $\alpha \in \Omega_B^r$ to the expression

$$(12.24) \qquad \gamma(\alpha) = (-1)^{(p+1)(r+1)+qr} * \alpha \wedge \varphi_0 \, .$$

Since $*\alpha \in F^{q-r}$ and $\varphi_0 \in F^2$, we have $\gamma(\alpha) \in F^{q-r+2} \Omega^{p+q-r+1}$, i.e. $\gamma(\alpha)$ is $\mathcal{F}$-trivial.

We can now compare δ_B with the usual $\delta : \Omega^r \to \Omega^{r-1}$ given by

$$\delta\alpha = (-1)^{n(r+1)+1} * d * \alpha$$

as follows. For $\alpha \in \Omega_B^r$

(12.25) $\delta\alpha = \delta_B\alpha + * \gamma(\alpha)$, where $* \gamma(\alpha)$ is orthogonal to $\Omega_B^{\cdot}$.

As a consequence, for $\alpha \in \Omega_B^r$, $\beta \in \Omega_B^{r-1}$

(12.26) $\langle \delta\alpha, \beta \rangle = \langle \delta_B\alpha, \beta \rangle_B$.

This identity proves that δ_B is the adjoint of d on basic forms. Further for $\alpha \in \Omega_B^r$ it follows by (12.25)

(12.27) $\gamma(\delta_B\alpha) = (-1)^{(n-r)r+1} * \delta * \gamma(\alpha) = (-1)^{n-r+1} d\gamma(\alpha)$.

These are the formulas which allow us to compare the basic Laplacian $\Delta_B = \delta_B d_B + d_B \delta_B$ and the ordinary Laplacian $\Delta = \delta d + d\delta$ restricted to basic forms. The result is as follows. For $\alpha \in \Omega_B^r$

(12.28) $\Delta\alpha = \Delta_B\alpha + \eta(\alpha)$, where

$$\eta(\alpha) = * \gamma(d_B\alpha) + d * \gamma(\alpha).$$

Note that $\eta(\alpha)$ is the restriction to $\alpha \in \Omega_B^r$ of the differential operator $\tilde{\eta} : \Omega^r \to \Omega^r$, of order one or less (and preserving degrees), given by

$$(12.29) \qquad \tilde{\eta}(\omega) = * \, \tilde{\gamma}(d\omega) + d * \tilde{\gamma}(\omega).$$

The content of (12.28) is that the differential operator $\Delta - \tilde{\eta} : \Omega^{\cdot} \to \Omega^{\cdot}$ is an extension of $\Delta_B : \Omega_B^{\cdot} \to \Omega_B^{\cdot}$. Since Δ is elliptic of order 2, and $\tilde{\eta}$ of lower order, $\Delta - \tilde{\eta}$ is still elliptic. The classical results applied to the elliptic operator $\Delta - \tilde{\eta}$ furnish the ingredients to conclude the desired results for the restriction Δ_B to Ω_B, for which the classical results do not directly apply. This is carried out in detail in [KT 16].

The first application of Theorem 12.22 is the unique representability of basic cohomology classes by basic harmonic forms. It is proved in the same way as the corresponding usual result in De Rham-Hodge Theory.

12.30 THEOREM. <u>Let the situation be as in Theorem 12.22. Then</u> $H_B^r(\mathcal{F}) \cong \mathcal{H}_B^r$.

PROOF. Let α be a closed basic r-form, and consider the De Rham-Hodge decomposition $\alpha = d_B\beta + \delta_B\gamma + \pi_B\alpha$, with $\pi_B : \Omega_B^r(\mathcal{F}) \to \mathcal{H}_B^r$ the orthogonal projection to harmonic forms. Then $0 = d_B\alpha = d_B\delta_B\gamma$ implies

$$\langle d_B\delta_B\gamma, \gamma\rangle_B = \langle \delta_B\gamma, \delta_B\gamma\rangle_B = \|\delta_B\gamma\|^2 = 0,$$

hence $\delta_B\gamma = 0$. It follows that $\alpha = d_B\beta + \pi_B\alpha$, and α is cohomologous to its harmonic representative $\pi_B\alpha$. If $\alpha = d_B\beta$, its harmonic representative vanishes. Thus we have a well-defined homomorphism $H_B^r(\mathcal{F}) \longrightarrow \mathcal{H}_B^r$, which is clearly surjective. If $\pi_B \alpha = 0$, then $\alpha = d_B \beta$, which proves that this map is also injective. ∎

Similarly we consider the De Rham-Hodge decomposition

(12.31) $$\Omega_B^r \cong \text{im } d_\kappa \oplus \text{im } \delta_\kappa \oplus \mathcal{H}_\kappa^r$$

with finite-dimensional $\mathcal{H}_\kappa^r = \ker \Delta_\kappa \subset \Omega_B^r$. The type of argument leading to Theorem 12.30 proves similarly that

(12.32) $$H_B^r(\mathcal{F}, d_\kappa) \equiv H^r(\Omega_B^{\cdot}(\mathcal{F}), d_\kappa) \cong \mathcal{H}_\kappa^r.$$

We use these results to prove the following fact [KT 11,13].

12.33 THEOREM (TWISTED DUALITY). <u>Let</u> $\mathcal{F}$ <u>be a transversally oriented Riemannian foliation, on a closed oriented manifold</u> M. <u>Assume</u> g_M <u>to be bundle-like with</u> $\kappa \in \Omega_B^1(\mathcal{F})$. <u>Then the pairing</u> $\alpha \otimes \beta \longrightarrow \int_M \alpha \wedge \beta \wedge \chi_\mathcal{F}$ <u>induces a nondegenerate pairing</u>

$$H_B^r(\mathcal{F}, d_B) \otimes H_B^{q-r}(\mathcal{F}, d_\kappa) \longrightarrow \mathbb{R}$$

<u>of finite-dimensional vector spaces</u>.

PROOF. Let $\alpha \in \Omega_B^r(\mathcal{F})$ with $d_B \alpha = 0$, and $\beta \in \Omega_B^{q-r}(\mathcal{F})$ with $d_\kappa \beta = 0$. Consider moreover $\alpha' = \alpha + d_B v$, and $\beta' = \beta + d_\kappa w$, with $v \in \Omega_B^{r-1}(\mathcal{F})$ and $w \in \Omega_B^{q-r-1}(\mathcal{F})$. Then

$$\alpha' \wedge \beta' \wedge \chi_{\mathcal{F}} = (\alpha + d_B v) \wedge (\beta + d_\kappa w) \wedge \chi_{\mathcal{F}}$$

$$= \alpha \wedge \beta \wedge \chi_{\mathcal{F}} + \alpha \wedge d_\kappa w \wedge \chi_{\mathcal{F}} + d_B v \wedge \beta' \wedge \chi_{\mathcal{F}}.$$

Now $d\alpha = d_B \alpha = 0$ implies

$$d(\alpha \wedge w \wedge \chi_{\mathcal{F}}) = (-1)^r \alpha \wedge dw \wedge \chi_{\mathcal{F}} + (-1)^{q-1} \alpha \wedge w \wedge d\chi_{\mathcal{F}}.$$

The last term differs according to (9.14) from $(-1)^{q-1} \alpha \wedge w \wedge (-\kappa \wedge \chi_{\mathcal{F}})$ by $(-1)^{q-1} \alpha \wedge w \wedge \varphi_0$, which is of filtration degree $r + (q - r - 1) + 2 = q + 1$, and hence vanishes. Thus

$$d(\alpha \wedge w \wedge \chi_{\mathcal{F}}) = (-1)^r \alpha \wedge dw \wedge \chi_{\mathcal{F}} + (-1)^q (-1)^{q-r-1} \alpha \wedge (\kappa \wedge w) \wedge \chi_{\mathcal{F}}$$

$$= (-1)^r \alpha \wedge (dw - \kappa \wedge w) \wedge \chi_{\mathcal{F}} = (-1)^r \alpha \wedge d_\kappa w \wedge \chi_{\mathcal{F}}.$$

Similarly $d_\kappa \beta = 0$ implies

$$d(v \wedge \beta' \wedge \chi_{\mathcal{F}}) = d_B v \wedge \beta' \wedge \chi_{\mathcal{F}} + (-1)^{r-1} v \wedge d\beta' \wedge \chi_{\mathcal{F}} + (-1)^{q-1} v \wedge \beta' \wedge d\chi_{\mathcal{F}},$$

where the last term can be replaced by $(-1)^{q-1} v \wedge \beta' \wedge (-\kappa \wedge \chi_{\mathcal{F}})$, since the difference $(-1)^{q-1} v \wedge \beta' \wedge \varphi_0$ is of filtration degree $(r - 1) + (q - r) + 2 = q + 1$, and hence vanishes. It follows that

$$d(v \wedge \beta' \wedge \chi_{\mathcal{F}}) = d_B v \wedge \beta' \wedge \chi_{\mathcal{F}} + (-1)^{r-1} v \wedge (d\beta' - \kappa \wedge \beta') \wedge \chi_{\mathcal{F}}$$

$$= d_B v \wedge \beta' \wedge \chi_{\mathcal{F}},$$

since $d_\kappa \beta' = d_\kappa \beta + d_\kappa^2 w = 0$. These calculations show that

(12.34) $\alpha' \wedge \beta' \wedge \chi_{\mathcal{F}} - \alpha \wedge \beta \wedge \chi_{\mathcal{F}} = (-1)^r d(\alpha \wedge w \wedge \chi_{\mathcal{F}}) + d(v \wedge \beta' \wedge \chi_{\mathcal{F}})$.

Therefore there is indeed a cohomology pairing as stated in the Theorem.

By Theorem 12.30 the cohomology spaces $H_B^r(\mathcal{F}, d_B)$ are finite-dimensional. (12.32) implies the same fact for the spaces $H_B^r(\mathcal{F}, d_\kappa)$. To complete the proof, it suffices therefore to establish the injectivity of the maps

$$H_B^r(\mathcal{F}, d_B) \longrightarrow H_B^{q-r}(\mathcal{F}, d_\kappa)^*, \quad H_B^{q-r}(\mathcal{F}, d_\kappa) \longrightarrow H_B^r(\mathcal{F}, d_B)^*$$

into the dual spaces defined by the pairing.

The first of these maps assigns to a d_B-closed basic r-form α the functional $[\beta] \longrightarrow \int_M \alpha \wedge \beta \wedge \chi_{\mathcal{F}}$, where $[\beta]$ is represented by a d_κ-closed (q - r)-form β. We can choose α to be Δ_B-harmonic, i.e. $\Delta_B \alpha = 0$. Now (12.21) implies that $\Delta_\kappa {}^* \alpha - {}^* \Delta_B \alpha = 0$, and ${}^* \alpha$ is Δ_κ-harmonic. It follows in particular that

$$\|\alpha\|^2 = \langle \alpha, \alpha \rangle_B = \int_M \alpha \wedge {}^* \alpha \wedge \chi_{\mathcal{F}} = 0,$$

and hence $\alpha = 0$. The injectivity of the other map is proved similarly. ∎

12.35 COROLLARY. <u>Let the situation be as in Theorem</u> 12.33. <u>Then</u> $H_B^q(\mathcal{F}, d_\kappa) \cong \mathbb{R}$. <u>Moreover, the following conditions are equivalent</u>:
(i) $\mathcal{F}$ <u>is taut</u>;
(ii) $H_B^q(\mathcal{F}) \cong \mathbb{R}$.

For $q = 1$, these conditions are always satisfied for a Riemannian foliation according to Theorem 7.43. On the other hand Carrière's example [CA 1,2] of Riemannian flows on 3-manifolds with $H_B^2(\mathcal{F}) = 0$ (see Chapter 10) shows that the alternative situation does occur.

For $q = n - 1$, and in the presence of a bundle-like metric g_M with $\kappa \in \Omega_B^1(\mathcal{F})$ this yields a proof of Theorem 10.17.

PROOF of 12.35. Since by (9.8) we have $H_B^0(\mathcal{F}, d_B) \cong \mathbb{R}$, Theorem 12.33 proves $H_B^q(\mathcal{F}, d_\kappa) \cong \mathbb{R}$. To prove (i) $\Rightarrow$ (ii), we observe that in the taut case there exists a bundle-like metric for which $\kappa = 0$. Thus $d_\kappa = d_B$ and $H_B^q(\mathcal{F}) \equiv H_B^q(\mathcal{F}, d_B) = \mathbb{R}$. To prove (ii) $\Rightarrow$ (i), assume g_M to be a bundle-like metric, such that $\kappa \in \Omega_B^1(\mathcal{F})$. Then the twisted duality implies that $H_B^0(\mathcal{F}, d_\kappa) \cong H_B^q(\mathcal{F}, d_B) \cong \mathbb{R}$. Thus there is a global nontrivial basic function $\lambda : M \longrightarrow \mathbb{R}$ satisfying

$$0 = d_\kappa \lambda \equiv d\lambda - \lambda \cdot \kappa.$$

It follows that $\kappa = d \log \lambda$. Since $\lambda \in \Omega_B^0(\mathcal{F})$, $[\kappa] = 0 \in H_B^1(\mathcal{F})$. We can now modify the metric g_M as in (12.7), so as to make the corresponding mean curvature form vanish. ∎

The argument just used shows that the taut case is characterized by $[\kappa] = 0 \in H_B^1(\mathcal{F})$. This is always the case for simply connected M.

12.36 COROLLARY. <u>Let the situation be as in Theorem</u> 12.33. <u>Then</u> $H_B^q(\mathcal{F}) \cong \mathbb{R}$ <u>or</u> $H_B^q(\mathcal{F}) = 0$. <u>The first case occurs if and only if</u> $\mathcal{F}$ <u>is taut</u>.

PROOF. The nontaut case occurs when $[\kappa] \neq 0 \in H_B^1(\mathcal{F})$. By the argument above this condition implies $H_B^0(\mathcal{F}, d_\kappa) = 0$. By Theorem 12.33 this implies $H_B^q(\mathcal{F}, d_B) = 0$. ∎

We finally formulate the particular case of Theorem 12.33 for a taut foliation [KT 9].

12.37 COROLLARY. Let $\mathcal{F}$ <u>be a taut and transversally oriented Riemannian foliation on a closed oriented manifold</u> M. <u>Then the pairing</u> $\alpha \otimes \beta \longrightarrow \int_M \alpha \wedge \beta \wedge \chi_\mathcal{F}$ <u>induces a nondegenerate pairing</u>

$$H_B^r(\mathcal{F}) \otimes H_B^{q-r}(\mathcal{F}) \longrightarrow \mathbb{R}$$

<u>on finite-dimensional vector spaces</u>.

Thus in the taut case Poincaré duality in $H_B^\cdot(\mathcal{F})$ holds in the expected form.

CHAPTER 13

A COMPARISON THEOREM

In this chapter we compare Riemannian foliations with transversally homogeneous foliations, where the model transverse structure is of the type of a compact symmetric space G/K. We state a comparison theorem [KRT 2] which is based on the results in Chapter 12.

We begin by describing transversally symmetric foliations. Let G/K be a Riemannian symmetric space of compact type with G and K connected, and q = dim G/K. The foliation $\mathcal{F}$ is transversally homogenous of type G/K, if $\mathcal{F}$ is given on an atlas of distinguished charts $U = \{U_\alpha\}$ by local submersions $f_\alpha : U_\alpha \to G/K$, related by transition functions given by the left action of an element $\gamma_{\alpha\beta} \in G : f_\alpha = \gamma_{\alpha\beta} f_\beta$ (see [BL 1]).

This transversal homogenity can be expressed in terms of the orthonormal frame bundle $F(Q)$ of Q as follows. The isotropy representation of G/K shows that $K \subset SO(q)$. Therefore, the transversal symmetric structure provides a K-reduction $K \to P \xrightarrow{\pi} M$ of $F(Q)$ with a foliated bundle structure [KT 3] [M 2]. This means that there is a K-invariant involutive subbundle $\tilde{L} \subset TP$, transversal to the fibers of P. The divided bundle $\tilde{L}/G$ on the base space M is the given $L \subset TM$. A connection on P is adapted to the foliated bundle structure, if the horizontal subspace contains $\tilde{L}$. Starting with a connection on P, the subspace $\tilde{L}_u$ is the horizontal lift of $L_{\pi(u)}$.

A $\underline{k}$-valued adapted connection η on P gives rise to a $\underline{g}$-valued Cartan connection

$$\omega = \eta + \varphi$$

where φ is the canonical $\mathbb{R}^q$-valued (solder) 1-form on P, defined by $\varphi(X) = u^{-1}(\pi(X))$, for $X \in T_u P$. The frame u of Q at $\pi(u)$ is viewed as a linear isomorphism $\mathbb{R}^q \to Q_{\pi(u)}$. The curvature

$$\Omega_\omega = d\omega + \frac{1}{2}[\omega,\omega]$$

can be expressed in terms of the curvature $\Omega_\eta = d\eta + \frac{1}{2}[\eta,\eta]$, and the torsion $\Phi_\eta = d\varphi + [\eta,\varphi]$, by

$$\Omega_\omega = \Omega_\eta + \frac{1}{2}[\varphi,\varphi] + \Phi_\eta,$$

where the brackets are expressed in terms of the brackets in the Lie algebra $\underline{g} = \underline{k} \oplus \underline{m}$.

In case η is the unique torsion-free metric connection, the symmetric space structure implies $\Omega_\eta = -\frac{1}{2}[\varphi,\varphi]$, and thus $\Omega_\omega = 0$. The last equation is the integrability condition for a locally symmetric transversal structure, and therefore is equivalent to the definition of a transversally symmetric foliation by local submersions outlined above.

We now compare Riemannian foliations with transversally symmetric foliations. An almost transversally symmetric foliation is one, where the Cartan curvature Ω_ω is small in an appropriate norm. In the spirit of Rauch's comparison theorem, and more specifically, the comparison theorem of Min-Oo and Ruh [MIR], one wishes to conclude that this assumption already implies the existence of a transversally symmetric structure of type G/K. One succeeds in doing so for Riemannian foliations with small basic mean curvature [KRT 2].

In the theorem below one allows a slightly more general situation. We start with a basic Cartan connection $\omega : TP \to \underline{g}$ with small curvature. It is not necessary to assume that the 1-form φ in $\omega = \eta + \varphi$, defined by the Cartan decomposition $\underline{g} = \underline{k} \oplus \underline{m}$, is the canonical 1-form. It suffices to assume that φ is nondegenerate. To simplify notations, we write Ω instead of Ω_ω. The Cartan connection ω is said to be adapted to the foliation $\mathcal{F}$ on P, if ω restricted to $\tilde{L}$ vanishes. An adapted Cartan connection is basic, if $i(\tilde{X})\Omega = 0$ for all $\tilde{X} \in \Gamma\tilde{L}$. This implies that Ω itself is a basic differential form. The following result is then proved in [KRT 2].

13.1 THEOREM. Let $\mathcal{F}$ be a transversally oriented Riemannian foliation of codimension $q \geq 2$ and basic mean curvature form κ, on the closed oriented Riemannian manifold (M, g_M). Let G/K be an irreducible compact symmetric space of dimension q and semi-simple $\underline{g}$. There exists a constant $A > 0$ depending only on the Lie algebra $\underline{g}$ and curvature bounds on M, with the following property. If $\omega : TP \to \underline{g}$ is a basic Cartan connection form, on the foliated K-reduction P of the normal frame bundle of $\mathcal{F}$, with Cartan curvature Ω and basic mean curvature form κ, then $\|\kappa\|_{1,\infty} + \|\Omega\|_{1,\infty} < A$ implies that $\mathcal{F}$ is transversally symmetric of type G/K.

As usual, $\|\Omega\|_{s,m}$ is the Sobolev norm of exponent m and involving the first s derivatives of Ω, and similarly for $\|\kappa\|_{s,m}$. For $m = \infty$ this is meant to indicate the essential supremum.

The idea of the proof is to construct a Cartan connection $\bar{\omega}$ with vanishing curvature. This yields a developing map $\Phi : \tilde{P} \to G$ on the universal covering $\tilde{P}$, equivariant with respect to a homomorphism $\pi_1(P) \longrightarrow \Gamma \subset G$ (holonomy of $\bar{\omega}$). It induces in turn a map $\varphi : \tilde{M} \to G/K$, possibly after an averaging process. This map defines the transversally symmetric structure of type G/K for the foliation $\mathcal{F}$, as asserted in the theorem, via its lift to the universal covering $\tilde{M}$.

One obtains $\bar{\omega} : TP \to \underline{g}$ as the limit of a sequence of Cartan connections. The sequence starts with $\omega^0 = \omega$, the Cartan connection of the Theorem. To define the iteration step, let $E = P \times \underline{g}$ denote the trivial vector bundle over P whose fiber is the Lie algebra $\underline{g}$. On E we define the linear connection

$$(13.2) \qquad D_X s = Xs + [\omega(X), s],$$

where s is a section in E, Xs is the derivative of s in direction X, $\omega = \omega^0$ is the original Cartan connection, and $[\,,\,]$ is the Lie bracket of $\underline{g}$. The curvature R^D of D is

$$(13.3) \qquad R^D(X,Y)s = [\Omega(X,Y), s].$$

In particular, R^D is a basic 2-form, in the basic complex $\Omega_B(\mathcal{F}, E)$ of E-valued differential forms on P for the foliation $\mathcal{F}$ (canonical lift of $\mathcal{F}$ to P).

The adjoint δ_B^E of d_B^E involves the mean curvature form $\tilde{\kappa}$ of $\mathcal{F}$. The Laplacian is as usual $\Delta_B^E = d_B^E \delta_B^E + \delta_B^E d_B^E$. Applying the DeRham-Hodge decomposition for Δ_B^E on $\Omega_B(\mathcal{F}, E)$ as discussed in Chapter 12, one defines

$$(13.4) \qquad \omega^{i+1} = \omega^i + \delta_B^E \beta^{i+1}.$$

Here β^{i+1} is the unique solution of

$$(13.5) \qquad \Delta_B^E \beta^{i+1} = -\Omega^i,$$

where Ω^i is curvature of ω^i. The uniqueness of β^{i+1} is a consequence of the nonexistence of harmonic forms in the complex $\Omega_B(\mathcal{F},E)$, which in turn follows from an estimate derived from a Bochner-Lichnerowicz formula.

Because the initial Cartan connection $\omega = \omega^o$, as well as the mean curvature form $\tilde{\kappa}$ of $\mathcal{F}$, are in the basic complex $\Omega_B(\mathcal{F},E)$, the connection forms ω^i and the curvature forms Ω^i are in this complex as well.

The idea is now that the sequence $\{\omega^i\}$ converges to a flat Cartan connection $\bar{\omega}$. The crucial result established in [KRT 2] is as follows.

13.6 CONVERGENCE LEMMA. <u>There exists a constant</u> $A' > 0$, <u>depending only on</u> g <u>and curvature bounds for the metric</u> g_M <u>on the basis</u> M <u>of</u> P, <u>such that</u> $(\|\kappa\|_{1,\infty} + \|\Omega\|_{1,\infty}) < A'$ <u>implies that the form</u> ω^{i+1} <u>of</u> (13.4) satisfies

(i) $\|\Omega^{i+1}\|_{1,m} < c(\|\kappa\|_{1,\infty} + \|\Omega\|_{1,\infty}) \|\Omega^i\|_{1,m}$,
(ii) $\|\omega^{i+1} - \omega^i\|_{2,m} < c\|\Omega^i\|_{1,m}$,

where c <u>is a constant depending only on</u> g <u>and curvature bounds on</u> g_M.

This is proved for exponents $m > \dim P$. The assertion (i) shows that $\{\|\Omega^i\|_{1,m}\}$ is a geometric sequence whose ratio can be made arbitrarily small by chosing A' suitably. This implies by assertion (ii), that $\sum_{i=0}^{\infty} \|\omega^{i+1} - \omega^i\|_{2,m}$ can be made arbitrarily small, by chosing A' suitably. Therefore $\{\omega^i\}$ converges to an element $\bar{\omega}$ in the Sobolev space $W_{2,m}$. Since $\|\omega - \bar{\omega}\|_{2,m}$ is small, $\bar{\omega}$ is nondegenerate, and hence a Cartan connection form. By (i), $\bar{\Omega} = d\bar{\omega} + \frac{1}{2}[\bar{\omega},\bar{\omega}] = 0$. The regularity theorem for this DE implies that $\bar{\omega}$ is a smooth differential form. This completes the outline of the proof of Theorem 13.1.

REFERENCES

[AM]   R. Almeida et P. Molino, Flots riemanniens sur les 4-variétés compactes, Tôhoku Math. J. 38(1986), 313-326.

[B]   M. Bauer, Feuilletage singulier défini par une distribution presque régulière, Thèse, Univ. de Strasbourg, 1985.

[BE]   M. Berger and D. Ebin, Some decompositions of the space of symmetric tensors on a Riemannian manifold, J. Diff. Geom. 3(1969), 376-392.

[BL1]   R. Blumenthal, Transversely homogeneous foliations, Ann. Institut Fourier 29(1979), 143-158.

[BL2]   R. Blumenthal, The base-like cohomology of a class of transversely homogeneous foliations, Bull. Sc. Math. 104(1980), 301-303.

[BL3]   R. Blumenthal, Riemannian homogeneous foliations without holonomy, Nagoya Math. J. 83(1981), 197-201.

[BL4]   R. Blumenthal, Riemannian foliations with parallel curvature, Nagoya Math. J. 90(1983), 145-153.

[BL5]   R. Blumenthal, Transverse curvature of foliated manifolds, Astérisque 116(1984), 25-30.

[BH]   R. Bott and A. Haefliger, On characteristic classes of Γ-foliations, Bull. Amer. Math. Soc. 78(1972), 1039-1044.

[BO1]   R. Bott, On a topological obstruction to integrability, Proc. Symp. in Pure Math., Amer. Math. Soc. 16(1970), 127-131.

[BO2]   R. Bott, Lectures on characteristic classes and foliations, Springer Lecture Notes in Math. 279(1972), 1-94.

[C]   C. Carathéodory, Untersuchungen über die Grundlagen der Thermodynamik, Math. Ann 67(1909), 355-386.

[C1]   G. Cairns, Géométrie globale des feuilletages totalement géodésiques, C. R. Acad. Sci. Paris 297(1983), 525-527.

[C2]   G. Cairns, Feuilletages totalement géodésiques de dimension 1, 2 ou 3. C. R. Acad. Sci. Paris 298(1984), 341-344.

[C3]   G. Cairns, Feuilletages totalement géodésiques, Séminaire de géométrie différentielle 1983-84, Montpellier.

[C4]   G. Cairns, Une remarque sur la cohomologie basique d'un feuilletage riemannien, Séminaire de géométrie différentielle 1984-85, Montpellier.

[CA1]   Y. Carrière, Flots riemanniens et feuilletages géodésibles de codimension un, Thèse Université des Sciences et Techniques de Lille I, 1981.

[CA2]   Y. Carrière, Flots riemanniens, Astérisque 116(1984), 31-52.

[CA3]   Y. Carrière, Les propriétés topologiques des flots riemanniens retrouvées à l'aide du théorème des variétés presque plates, Math. Z. 186(1984), 393-400.

[CA4]   Y. Carrière, Sur la croissance des feuilletages de Lie, preprint.

[CA5]   Y. Carrière, Feuilletages riemanniens a croissance polynomiale, preprint.

[CAC]   P. Caron et Y. Carrière, Flots transversalements de Lie $\mathbb{R}^n$, flots de Lie minimaux, C. R. Acad. Sc. Paris 280(1980), 477-478.

[CAGH] Y. Carrière et E. Ghys, Feuilletages totalement géodésiques, Anais da Acad. Bras. de Ciencias 53(1981), 427-432.

[CC]   J. Cantwell and L. Conlon, The dynamics of open, foliated manifolds and a vanishing theorem for the Godbillon-Vey class, Advances in Math. 53(1984), 1-27.

[CL]   E. Calabi, An intrinsic characterization of harmonic 1-forms, Global Analysis, Papers in honor of K. Kodaira, Ed. by D. C. Spencer and S. Iyanaga, Princeton Math. Series 29(1969), 101-117.

[DI]   P. Dirac, Quantised singularities in the electromagnetic field, Proc. Royal Soc. of London A 113(1931), 60-71.

[D]   T. Duchamp, Characteristic invariants of G-foliations, Ph.D. Thesis, University of Illinois, Urbana, Illinois, 1976.

[E]   J. Eells, Elliptic operators on manifolds, complex analysis and its appl. (Trieste), Vol. I(1975), 95-152.

[EH1]   A. El Kacimi-Alaoui et G. Hector, Décomposition de Hodge sur l'espace des feuilles d'un feuilletage riemannien, C. R. Acad. Sci. Paris 298(1984), 289-292.

[EH2]   A. El Kacimi-Alaoui et G. Hector, Decomposition de Hodge basique pour un feuilletage riemannien, Ann. Inst. Fourier 36(1986), 207-227.

[ER]   C. Ehresmann et G. Reeb, Sur les champs d'éléments de contact de dimension p complètement intégrables dans une variété continuement différentiable V_n, C.R. Acad. Sc. Paris 216(1944), 628-630.

[ES]   J. Eells and J. H. Sampson, Harmonic mappings of riemannian manifolds, Amer. J. Math. 86(1964), 109-160.

[ESH] A. El Kacimi-Alaoui, V. Sergiescu et G. Hector, La cohomologie basique d'un feuilletage riemannien est de dimension finie, Math. Z. 188(1985), 593-599.

[F] E. Fedida, Sur les feuilletages de Lie, C. R. Acad. Sc. Paris 272(1971), 999-1002.

[GH1] E. Ghys, Classification des feuilletages totalement géodésiques de codimension un, Comment. Math. Helv. 58(1983), 543-572.

[GH2] E. Ghys, Feuilletages riemanniens sur les variétés simplement connexes, Ann. Inst. Fourier 34(1984), 203-223.

[GH3] E. Ghys, Groupes d'holonomie des feuilletages de Lie, Proc. Kon, Nederl. Acad. Sci. A 88(1985), 173-182.

[GH4] E. Ghys, Un feuilletage analytique dont la cohomologie basique est de dimension infinie, to appear.

[GHS] E. Ghys et V. Sergiescu, Stabilité et conjugaison différentiable pour certains feuilletages, Topology 19(1980), 179-197.

[GHV] W. Greub, S. Halperin and R. Vanstone, Connections, curvature and cohomology, Vol. II (1973), Academic Press.

[GL1] H. Gluck, Can space be filled by geodesics, and if so, how?, open letter, 1979.

[GL2] H. Gluck, Dynamical behavior of geodesic fields, Springer Lecture Notes in Math. 819(1980), 190-215.

[GS] V. Guillemin and S. Sternberg, Geometric asymptotics, Math. Surveys 14(1977), Amer. Math. Soc.

[GV] C. Godbillon and J. Vey, Un invariant des feuilletages de codimension un, C. R. Acad. Sci. Paris 273(1971), 92-95.

[HA1] A. Haefliger, Structures feuilletées et cohomologie a valeur dans un faisceau de groupoides, Comment. Math. Helv. 32(1958), 249-329.

[HA2] A. Haefliger, Variétés feuilletées, Ann. Scuola Norm. Sup. Pisa 16(1962), 249-329.

[HA3] A. Haefliger, Feuilletages sur les variétés ouvertes, Topology 9(1970), 183-194.

[HA4] A. Haefliger, Sur les classes caractéristiques des feuilletages, Sém. Bourbaki 412-01 to 412-21 (1972), Springer Lecture Notes in Math. 317(1973).

[HA5] A. Haefliger, Differentiable cohomology, C.I.M.E. Lectures 1976.

[HA6] A. Haefliger, Some remarks on foliations with minimal leaves, J. Diff. Geom. 15(1980), 269-284.

[HA7]   A. Haefliger, Groupoïdes d'holonomie et classifiants, Astérisque 116(1984), 70-97.

[HA8]   A. Haefliger, Pseudogroups of local isometries, Pitman Research Notes 131(1985), 174-197.

[HB]   J. Hebda, Curvature and focal points in Riemannian foliations, Indiana Univ. Math. J. 35(1986), 321-331.

[HE1]   R. Hermann, A sufficient condition that a map of riemannian manifolds be a fiber bundle, Proc. Amer. Math. Soc. 11(1960), 236-242.

[HE2]   R. Hermann, The differential geometry of foliations; I, Ann. of Math. 72(1961), 445-457; II, J. Math. and Mech. 11(1962), 303-316.

[HH]   J. Heitsch and S. Hurder, Secondary classes, Weil measures and the geometry of foliations, J. Diff. Geom. 20(1984), 291-309.

[HKA]   S. Hurder and F. Kamber, Homotopy invariants of foliations, Topology Siegen 1979, Springer Lecture Notes in Math. 788(1980), 49-61.

[HK1]   S. Hurder and A. Katok, Secondary classes and transverse measure theory of a foliation, Bull. Amer. Math. Soc. 11(1984), 347-350.

[HK2]   S. Hurder and A. Katok, Smoothness and Godbillon-Vey classes of geodesic horocycle foliation on surfaces of variable negative curvature, to appear.

[HL]   R. Harvey and H. B. Lawson, Calibrated foliations, Amer. J. of Math. 104(1982), 607-633.

[HT]   M. Hvidsten and Ph. Tondeur, A characterization of harmonic foliations by variations of the metric, Proc. Amer. Math. Soc. 98(1986), 359-362.

[HU1]   S. Hurder, Dual homotopy invariants of G-foliations, Topology 20(1981), 365-387.

[HU2]   S. Hurder, On the homotopy and cohomology of the classifying space of Riemannian foliations, Proc. Amer. Math. Soc. 81(1981), 484-489.

[HU3]   S. Hurder, On the secondary classes of foliations with trivial normal bundles, Comment. Math. Helv. 56(1981), 307-326.

[HU4]   S. Hurder, Independent rigid secondary classes for holomorphic foliations, Invent. Math. 66(1982), 313-323.

[HU5]   S. Hurder, Vanishing of secondary classes for compact foliations, J. London Math. Soc. 28(1983), 175-183.

[HU6]   S. Hurder, The classifying space of smooth foliations, Ill. J. of Math. 29(1985), 108-133.

[HU7]   S. Hurder, Growth of leaves and secondary invariants of foliations, preprint.

[HU8]   S. Hurder, Exotic classes for measured foliations, Bull. Amer. Math. Soc. 7(1982), 389-391.

[HU9]   S. Hurder, Global invariant for measured foliations, Trans. Amer. Math. Soc. 280(1983), 367-391.

[HU10] S. Hurder, The Godbillon measure for amenable foliations, preprint.

[J]   R. Jerrard, Curvatures of surfaces associated with holomorphic functions, Colloq. Math. 21(1970), 127-132.

[JN]   D. Johnson and A. Naveira, A topological obstruction to the geodesibility of a foliation of odd dimension, Geom. Dedicata 11(1981), 347-357.

[JR]   R. Jerrard and L. Rubel, On the curvature of the level lines of a harmonic function, Proc. Amer. Math. Soc. 14(1963), 29-32.

[JW1]   D. Johnson and L. Whitt, Totally geodesic foliations on 3-manifolds, Proc. Amer. Math. Soc. 76(1979), 355-357.

[JW2]   D. Johnson and L. Whitt, Totally geodesic foliations, J. Diff. Geom. 15(1980), 225-235.

[KN]   S. Kobayashi and K. Nomizu, Foundations of differential geometry I, II(1963, 1969).

[KO]   S. Kobayashi, Transformation groups in differential geometry, Ergeb. Math. 70(1972), Springer Verlag, New York.

[KRT1] F. Kamber, E. Ruh and Ph. Tondeur, Almost transversally symmetric foliations, Proc. of the II. Int. Symp. on differential geometry, Peniscola 1985, Springer Lecture Notes 1209(1986), 184-189.

[KRT2] F. Kamber, E. Ruh and Ph. Tondeur, Comparing Riemannian foliations with transversally symmetric foliations, J. of Diff. Geom. 27(1988).

[KT1]   F. Kamber and Ph. Tondeur, Invariant differential operators and the cohomology of Lie algebra sheaves, Memoirs Amer. Math. Soc. 113(1971), 1-125.

[KT2]   F. Kamber and Ph. Tondeur, Characteristic invariants of foliated bundles, Manuscripta Math. 11(1974), 51-89.

[KT3]   F. Kamber and Ph. Tondeur, Foliated bundles and characteristic classes, Springer Lecture Notes in Math. 494(1975).

[KT4]   F. Kamber and Ph. Tondeur, G-foliations and their characteristic classes, Bull. Amer. Math. Soc. 84(1978), 1086-1124.

[KT5]   F. Kamber and Ph. Tondeur, Feuilletages harmoniques, C. R. Acad. Sci. Paris 291(1980), 409-411.

[KT6]   F. Kamber and Ph. Tondeur, Harmonic foliations Proc. NSF Conference on Harmonic Maps, Tulane (1980), Springer Lecture Notes in Math. 949(1982), 87-121.

[KT7]   F. Kamber and Ph. Tondeur, Infinitesimal automorphisms and second variation of the energy for harmonic foliations, Tôhoku Math. J. 34(1982), 525-538.

[KT8]   F. Kamber and Ph. Tondeur, Dualité de Poincaré pour les feuilletages harmoniques, C. R. Acad Sci. Paris 294(1982), 357-359.

[KT9]   F. Kamber and Ph. Tondeur, Duality for riemannian foliations, Proc. Symp. Pure Math. Vol. 40(1983); Part 1, 609-618.

[KT10]   F. Kamber and Ph. Tondeur, The index of harmonic foliations on spheres, Trans. Amer. Math. Soc. 275(1983), 257-263.

[KT11]   F. Kamber and Ph. Tondeur, Foliations and metrics, Proc. of a year in Differential Geometry, University of Maryland, Birkhäuser, Progress in Mathematics Vol. 32(1983), 103-152.

[KT12]   F. Kamber and Ph. Tondeur, Curvature properties of harmonic foliations, Illinois J. of Math. 18(1984), 458-471.

[KT13]   F. Kamber and Ph. Tondeur, Duality theorems for foliations, Astérisque 116(1984), 108-116.

[KT14]   F. Kamber and Ph. Tondeur, The Bernstein problem for foliations, Proc. of the Conference on Global Differential Geometry and Global Analysis, Berlin 1984, Springer Lecture Notes in Math. 1156(1985), 216-218.

[KT15]   F. Kamber and Ph. Tondeur, Foliations and harmonic forms, Colloque sur les applications harmoniques, Luminy 1986, to appear.

[KT16]   F. Kamber and Ph. Tondeur, De Rham-Hodge theory for riemannian foliations, Math. Ann. 277(1987), 415-431.

[KTT]   F. Kamber, Ph. Tondeur and G. Toth, Transversal Jacobi fields for harmonic foliations, Michigan Math. J. 34(1987), 261-266.

[L1]   H. Lawson, Foliations, Bull. Amer. Math. Soc. 80(1974), 369-418.

[L2]   H. Lawson, Lectures on the quantitative theory of foliation, CBMS Regional Conf. Series, Vol. 27(1977).

[LB]   F. Laudenbach et S. Blank, Isotopie de formes fermées en dimension trois, Invent. Math. 54(1979), 103-177.

[LN1]   R. Langevin, Feuilletages tendus, Bull. Soc. Math. France 107(1979), 271-281.

[LN2] R. Langevin, Courbure, feuilletages et surfaces (mesures et distributions de Gauss), Publ. Math. d'Orsay no 80-83 (1980).

[LN3] R. Langevin, Energies et géométrie intégrale, I. Int. Symp. on differential geometry, Peniscola 1982, Springer Lecture Notes 1045(1984), 95-103.

[LN4] R. Langevin, Feuilletages, énergies et cristaux liquides, Astérisque 107-108(1983), 201-213.

[LP] C. Lazarov and J. Pasternack, Secondary characteristic classes for Riemannian foliations, J. Diff. Geom 11(1976), 365-385.

[M1] P. Molino, Connexions et G-structures sur les variétés feuilletées, Bull. Sc. Math. 92(1968), 59-63.

[M2] P. Molino, Classe d'Atiyah d'un feuilletage et connexions transverses projetables, C. R. Acad. Sc. Paris 272(1971), 779-781.

[M3] P. Molino, Classes caractéristiques et obstructions d'Atiyah pour les fibrés principaux feuilletés, C. R. Acad. Sc. Paris 272(1971), 1376-1378.

[M4] P. Molino, Propriétés cohomologiques et propriétés topologiques des feuilletages à connexions transverses projetables, Topology 12(1973), 317-325.

[M5] P. Molino, Feuilletages transversement parallélisables et feuilletages de Lie, C. R. Acad. Sc. Paris 282(1976), 99-101.

[M6] P. Molino, Feuilletages transversalement complets et applications, Ann. Ec. Norm. Sup. Paris 70(1977), 289-307.

[M7] P. Molino, Feuilletages riemanniens sur les variétés compactes; champs de Killing transverses, C. R. Acad. Sc. Paris 289(1979), 421-423.

[M8] P. Molino, Géometrie globale des feuilletages riemanniens, Proc. Kon. Nederland Akad. Ser. A, 1, 85(1982), 45-76.

[M9] P. Molino, Feuilletages riemanniens, Secrétariat des Mathématiques, Université des Sciences et Technique du Languedoc, 1982-1983.

[M10] P. Molino, Flots riemanniens et flots isométriques, Séminaire de géometrie différentielle 1982-83, Montpellier.

[M11] P. Molino, Feuilletages riemanniens réguliers et singuliers, to appear.

[MIR] Min-Oo and E. Ruh, Comparison theorems for compact symmetric space, Ann. Scient. Ec. Norm Sup., 4e série, 12(1979), 335-353.

[MS] P. Molino et V. Sergiescu, Deux remarques sur les flots riemanniens, Manuscripta Math. 51(1985), 145-161.

[MR] J. Moser, On the volume elements on a manifold, Trans. Amer. Math. Soc. 120(1965), 286-294.

[MZ1] H. Münzner, Isoparametrische Hyperflächen in Sphären, Math. Ann. 251(1980), 57-71.

[MZ2] H. Münzner, Isoparametrische Hyperflächen in Sphären, II, Über die Zerlegung der Sphäre in Ballbündel, Math. Ann. 256(1981), 215-232.

[N] S. Novikov, Topology of foliations, Trudy Moskov. Mat. Obsc. 14(1965), 248-278; AMS Translation, Trans. Moscow Math. Soc. 14(1967), 268-304.

[NT] S. Nishikawa and Ph. Tondeur, Transversal infinitesimal automorphisms for harmonic Kähler foliations, to appear.

[NZ] K. Nomizu, Some results in E. Cartan's theory of isoparametric families of hypersurfaces, Bull. Amer. Math. Soc. 79(1973), 1184-1188.

[ON] B. O'Neill, The fundamental equations of a submersion, Michigan Math. J. 13(1966), 459-469.

[OS1] G. Oshikiri, A remark on minimal foliations, Tôhoku Math. J. 33(1981), 133-137.

[OS2] G. Oshikiri, Totally geodesic foliations and Killing fields, Tôhoku Math. J. 35(1983), 387-392.

[OS3] G. Oshikiri, Totally geodesic foliations and Killing fields, II, Tôhoko Math. J. 38(1986), 351-356.

[PA] J. Pasternack, Foliations and compact group actions, Comment. Math. Helv. 46(1971), 467-477.

[PL] J. Plante, Foliations with measure preserving holonomy, Ann. of Math. 102(1975), 327-361.

[PO] W. Poor, Differential geometric structures, McGraw-Hill, New York (1981).

[R1] G. Reeb, Sur certaines propriétés topologiques des variétés feuilletées, Actualités Sci. Indust. Hermann, Paris (1952).

[R2] G. Reeb, Structures feuilletées, Springer Lecture Notes in Math. 652(1978), 104-113.

[RE1] B. Reinhart, Harmonic integrals on almost product manifolds, Trans. Amer. Math. Soc. 88(1958), 243-276.

[RE2] B. Reinhart, Foliated manifolds with bundle-like metrics, Ann. of Math. 69(1959), 119-132.

[RE3] B. Reinhart, Harmonic integrals on foliated manifolds, Amer. J. of Math. 81(1959), 529-536.

[RE4]   B. Reinhart, Closed metric foliations, Mich. Math. J. 8(1961), 7-9.

[RE5]   B. Reinhart, Structures transverse to a vector field, Int. Symp. on nonlinear differential equations and nonlinear mechanics, Academic Press, New York, 1963, 442-444.

[RE6]   B. Reinhart, A metric formula for the Godbillon-Vey invariant for foliations, Proc. Amer. Math. Soc. 38(1973), 427-430.

[RE7]   B. Reinhart, The second fundamental form of a plane field, J. Diff. Geom. 12(1977), 619-627.

[RE8]   B. Reinhart, Foliations and second fundamental form, Fourth Colloquium on differential geometry, Santiago de Compostela, 1978, 246-253.

[RE9]   B. Reinhart, Differential geometry of foliations, Ergeb. Math. 99(1983), Springer-Verlag, New York.

[RU1]   H. Rummler, Quelques notions simples en géométrie riemannienne et leurs applications aux feuilletages compacts, Comment. Math. Helv. 54(1979), 224-239.

[RU2]   H. Rummler, Kompakte Blätterungen durch Minimalflächen, Habilitations-schrift Universität Freiburg i.Ue. (1979).

[S]   R. Sacksteder, Foliations and pseudogroups, Amer. J. of Math. 87(1965), 79-102.

[SA]   M. Saralegui, The Euler class for flows of isometries, Pitman Research Notes in Math. 131(1986), 220-227.

[SCH]   G. Schwarz, On the De Rham cohomology of the leaf space of a foliation, Topology 13(1974), 185-187.

[SE]   V. Sergiescu, Cohomologie basique et dualité des feuilletages riemanniens, Ann. Inst. Fourier 35(1985), 137-158.

[SI]   J. Sikorav, Formes différentielles fermées sur le n-tore, Comment. Math. Helv. 57(1982), 79-106.

[SU1]   D. Sullivan, Cycles for the dynamical study of foliated manifolds and complex manifolds, Invent. Math. 36(1976), 225-255.

[SU2]   D. Sullivan, A foliation of geodesics is characterized by having no tangent homologies, J. Pure Appl. Algebra 13(1978), 101-104.

[SU3]   D. Sullivan, A homological characterization of foliations consisting of minimal surfaces, Comment. Math. Helv. 54(1979), 218-223.

[T]   Ph. Tondeur, The mean curvature of Riemannian foliations, Colloque de géométrie symplectique et physique mathématique, Lyon 1986, to appear.

[TH1] W. Thurston, Non-cobordant foliations on S^3, Bull. Amer. Math. Soc. 78(1972), 511-514.

[TH2] W. Thurston, The theory of foliations of codimension greater than one, Comment. Math. Helv. 49(1974), 214-231.

[TH3] W. Thurston, A generalization of the Reeb stability theorem, Topology 13(1974), 347-352.

[TH4] W. Thurston, Foliations and groups of diffeomorphisms, Bull. Amer. Math. Soc. 80(1974), 304-307.

[TH5] W. Thurston, Existence of codimension one foliations, Ann. of Math. 104(1976), 249-268.

[TH6] W. Thurston, Hyperbolic structures on 3-manifolds, II: Surface groups and 3-manifolds which fiber over the circle, Ann. of Math., to appear.

[TI] D. Tischler, On fibering certain foliated manifolds over S^1, Topology 9(1970), 153-154.

[TT] Ph. Tondeur and G. Toth, On transversal infinitesimal automorphisms for harmonic foliations, Geometriae Dedicata 24(1987), 229-236.

[W] A. Wadsley, Geodesic foliations by circles, J. Diff. Geom. 10(1975), 541-549.

[WU] H. Wu, A remark on the Bochner technique in differential geometry, Proc. Amer. Math. Soc. 78(1980), 403-408.

APPENDIX: BIBLIOGRAPHY ON FOLIATIONS

K. Abe, Characterization of totally geodesic submanifolds of S^N and CP^N by an inequality, Tôhoku Math. J. 23(1971), 219-244.

K. Abe, Applications of a Ricatti type differential equation to Riemannian manifolds with totally geodesic distributions, Tôhoku Math. J. 25(1973), 425-444.

N. Abe, On foliations and exotic characteristic classes, Kodai Math. Sem. Rep. 28(1977), 324-341.

N. Abe, Exotic characteristic classes of certain Γ-foliations, Kodai Math. J. 2(1979), 254-271.

Y. Abe, On Levi foliations, Memoirs of the Faculty of Science, Kyushu Univ., Ser. A 38(1984), 169-176.

N. A'Campo, Un feuilletage de S^5, C. R. Acad. Sci. 272(1971), 1504-1506.

N. A'Campo, Feuilletages de codimension 1 sur des variétés de dimension 5, C. R. Acad. Sci. 273(1971), 603-604.

N. A'Campo, Feuilletages de codimension 1 sur les variétés simplement convexes de dimension 5, Comment. Math. Helv. 47(1972), 514-525.

Y. Agaoka, Geometric invariants associated with flat projective structures, J. Math. Kyoto Univ. 22(1983), 701-718.

C. Albert, Feuilletages invariants et pseudoalgebres de Lie lisses, Cah. Topol. Geom. Diff., Ch. Ehresmann 13(1972), 309-323.

C. Albert, Invariants riemanniens des champs de plans, C. R. Acad. Sci. Paris 296(1983), 329-332.

C. Albert and D. Lehmann, Une algèbre graduée universelle pour les connexions sans torsion, Math. Z. 159(1978), 133-142.

R. Almeida and P. Molino, Flots riemanniens sur les 4-variétés compactes, Tôhoku Math. J. 38(1986), 313-326.

R. Almeida and P. Molino, Suites d'Atiyah et feuilletages transversalement complets, C. R. Acad. Sci. Paris 300(1985), 13-15.

J. Alvarez, A finiteness theorem for the spectral sequence of a Riemannian foliation, Ill. J. of Math., to appear.

J. Alvarez, Duality in the spectral sequence of a Riemannian foliation, preprint.

G. Andrzejczak, Some characteristic invariants of foliated bundles, Dissertationes Math. (Rozprawy Mat.) 222(1984), 67pp.

G. Andrzejczak, More characteristic invariants of foliated bundles, Diff. Geom., Banach Center Publ. 12(1984), 9-27.

M. Anona, Sur la d_L-cohomologie, C. R. Acad. Sci. Paris 290(1980), 649-651.

D. Anosov, Roughness of geodesic flows on compact Riemannian manifolds of negative curvature, Dokl. Akad. Nauk SSSR 145, 707-709 [Russian]. Translation: Soviet Math. Dokl. 3(1962), 1068-1069.

D. Anosov, Erogodic properties of geodesic flows on closed Riemannian manifolds of negative curvature, Dokl. Akad. Nauk SSSR 151, 1250-1252 [Russian]. Translation: Soviet Math. Dokl. 4(1963), 1153-1156.

D. Anosov, Geodesic flows on closed Riemannian manifolds with negative curvature, Trudy Mat. Inst. Steklov 90 [Russian]. Translation: Proc. Steklov Inst., Math. 90(1967).

L. Apostolova, Nearly Kähler manifolds are holomorphic foliations, C. R. Acad. Bulgare Sci. 38(1985), 977-979.

C. Apreutesei, Quelques classes caractéristiques et G_T-structure, C. R. Acad. Sci. Paris 280(1975), 41-44.

C. Apreutesei, Algèbres HC(;), HP(;) et obstructions à l'intégrabilité, Att Accad. Naz. Lincei. Rend. Cl. Sci. Fis. Mat. Nat. 62(1977), 17-25.

J. Arraut, A two-dimensional foliation on S^7, Topology 3(1973), 243-245.

D. Asimov, Round handles and homotopy of nonsingular vector fields, Bull. Amer. Math. Soc. 81(1975), 417-419.

D. Asimov, On the average Gaussian curvature of leaves of foliations, Bull. Amer. Math. Soc. 84(1978), 131-133.

D. Asimov, and H. Gluck, Morse-Smale fields of geodesics, In: Global Theory of Dynamical Systems, Springer Lecture Notes in Math. 819(1980), 1-17.

M. Atiyah, Complex analytic connections in fibre bundles, Trans. Amer. Math. Soc. 85(1957), 181-207.

M. Atiyah, Vector fields on manifolds, Arbeitsgemeinschaft für Forschung des Landes Nordrhein-Westfalen, vol. 200, Westdeutscher Verlag (1970).

M. Atiyah, Elliptic operators and compact groups, Springer Lecture Notes in Math. 401(1974).

D. Baker, On a class of foliations and the evaluation of their characteristic classes, Bull. Amer. Math. Soc. 83(1977), 394-396.

D. Baker, On a class of foliations and the evaluation of their characteristic classes, Comment. Math. Helv. 53(1978), 334-363.

D. Baker, Some cohomology invariants for deformations of foliations, Ill. J. of Math. 25(1981), 169-189.

M. Bauer, Almost regular foliations, C. R. Acad. Sci. Paris 299(1984), 387-390.

M. Bauer, Codimension one, almost regular foliations, C. R. Acad. Sci. Paris 299(1984), 819-822.

M. Bauer, Feuilletage singulier défini par une distribution presque régulière, Thèse, Univ. de Strasbourg (1985).

P. Baum, Structure of foliation singularities, Adv. Math. 15(1975), 361-374.

P. Baum and R. Bott, On the zeros of meromorphic vectorfields, Essays Topol. Relat. Topics, Berlin et al (1970), 29-47.

P. Baum and R. Bott, Singularities of holomorphic foliations, J. Diff. Geometry 7(1972), 279-342.

E. Bedford and M. Kalka, Foliations and Monge-Ampère equations, Comm. Pure Appl. Math. 30(1977), 543-571.

I. Belko, Affine transformations of a transversal projectable connection on a manifold with a foliation. Mat. Sbor. 117(1982), 181-195; Math. USSR Sbornik 45(1983), 191-204.

S. Benenti and W. Tulczyjew, Sur un feuilletage coisotrope du fibré cotangent d'un groupe de Lie, C. R. Acad. Sci. Paris 300(1985), 119-122.

C. Benson, Characteristic classes for symplectic foliations, Mich. Math. J. 33(1986), 105-118.

C. Benson and R. Ellis, Characteristic classes of transversely homogeneous foliations, Trans. Amer. Math. Soc. 289(1985), 849-859.

M. Berger and D. Ebin, Some decompositions of the space of symmetric tensors on a Riemannian manifold, J. Diff. Geom. 3(1969), 376-392.

I. Bernshtein, and B. Rozenfeld, On characteristic classes of foliations, Funkcional. Anal. i Prilozen 6, 68-69 [Russian]. Translation: Functional Anal. Apppl. 6(1972), 60-61.

I. Bernshtein, and B. Rozenfeld, Homogeneous spaces of infinite-dimensional Lie algebras and characteristic classes of foliations. Uspehi Mat. Nauk. 28, 103-138 [Russian]. Translation: Russian Math. Surveys 28(1973), 107-142.

I. Bivens, Orthogonal geodesics and minimal distributions, Trans. Amer. Math. Soc. 275(1983), 397-408.

D. Blair and J. Vanstone, A generalization of the helicoid, Minimal submanifolds and geodesics, Proc. Japan-US Seminar Tokyo 1977, 13-16.

R. Blumenthal, Transversely homogeneous foliations, Ann. Institut Fourier 29(1979), 143-158.

R. Blumenthal, The base-like cohomology of a class of transversely homogeneous foliations, Bull. des Sciences Math. 104(1980), 301-303.

R. Blumenthal, Riemannian homogeneous foliations without holonomy, Nagoya Math. J. 83(1981), 197-207.

R. Blumenthal, Foliated manifolds with flat basic connection, J. Diff. Geom. 16(1981), 401-406.

R. Blumenthal, Basic connections with vanishing curvature and parallel torsion, Bull. des Sciences Math. 106(1982), 393-400.

R. Blumenthal, Riemannian foliations with parallel curvature, Nagoya Math. J. 90(1983), 145-153.

R. Blumenthal, Transverse curvature of foliated manifolds, Astérisque 116(1984), 25-30.

R. Blumenthal, Foliations with locally reductive normal bundle, Ill. J. Math. 28(1984), 691-702.

R. Blumenthal, Stability theorems for conformal foliations, Proc. Amer. Math. Soc. 91(1984), 485-491.

R. Blumenthal, Cartan connections in foliated bundles, Mich. Math. J. 31(1984), 55-63.

R. Blumenthal, Local isomorphisms of projective and conformal structures, Geometriae Dedicata 16(1984), 73-78.

R. Blumenthal, Affine submersions, and foliations of affinely connected manifolds, C. R. Acad. Sci. Paris 299(1984), 1013-1015.

R. Blumenthal, Connections on foliated manifolds, Springer Lecture Notes 1165(1985), 30-35.

R. Blumenthal, Sprays, fiber spaces, and product decompositions, Proceedings of the Fifth International Colloquium on Differential Geometry, Santiago de Compostela (1984), Pitman Research Notes 131(1985), 156-161.

R. Blumenthal, Affine submersions, Ann. Global Anal. Geom. 3(1985), 275-285.

R. Blumenthal, Cartan submersions and Cartan foliations, Ill. J. Math. 31(1987), 327-343.

R. Blumenthal, Les applications de Cartan et les feuilletages à modèle transverse un espace à connexion de Cartan, C. R. Acad. Sci. Paris 301(1985), 919-922.

R. Blumenthal, Mappings between manifolds with Cartan connections, Springer Lecture Notes 1209(1986), 94-99.

R. Blumenthal and J. Hebda, Complementary distributions which preserve the leaf geometry and applications to totally geodesic foliations, Quarterly J. Math. Oxford 35(1984), 383-392.

R. Blumenthal and J. Hebda, De Rham decomposition theorems for foliated manifolds, Ann. Inst. Fourier 33(1983), 183-198.

R. Blumenthal and J. Hebda, Ehresmann connections for foliations, Indiana Univ. Math. J. 33(1984), 597-611.

R. Blumenthal and J. Hebda, Un analogue de la nappe d'holonomie pour une variété feuilletée, C. R. Acad. Sci. Paris 303(1986), 931-934.

R. Blumenthal and J. Hebda, An analogue of the holonomy bundle for a foliated manifold, Tôhoku Math. J., to appear.

C. Bonatti, Sur les feuilletages singuliers stables des variétés de dimension trois, Comment. Math. Helv. 60(1985), 429-444.

C. Bonatti, Existence of codimension one singular foliations with dense leaves on closed manifolds, C. R. Acad. Sci. Paris 300(1985), 493-496.

A. Bonome and L. A. Cordero, The GLA-cohomology of vector-valued differential forms on foliated manifolds, Boletin Acad. Galega de Ciencias, vol. I(1982), 53-65.

W. Boothby, Transversely complete e-foliations of codimension one and accessibility properties of non-linear systems, Lie groups: History, Frontiers and Appl. Vol. VII, Math. Sci. Press, Brookline (1977), 361-385.

R. Bott, Vector fields and characteristic numbers, Mich. Math. J. 14(1967), 231-244.

R. Bott, On a topological obstruction to integrability. In: Global Analysis, Proceedings of Symposia in Pure Math., vol. 16(1970), 127-131.

R. Bott, On topological obstructions to integrability, Actes Congr. Int. Mathematiciens 1970, 1, Paris (1971), 27-36.

R. Bott, Lectures on characteristic classes and foliations, Springer Lecture Notes in Math. 279(1972), 1-94.

R. Bott, On the Lefschetz formula and exotic characteristic classes, Symp. Math. Conv., 1971-1972, London-New York, 10(1972), 95-105.

R. Bott, Gelfand-Fuks cohomology and foliations, Proc. Symp. New Mexico State University (1973).

R. Bott, On characteristic classes in the framework of Gelfand-Fuks cohomology, Astérisque 32-33(1976), 113-139.

R. Bott, On the Chern-Weil homomorphism and the continuous cohomology of Lie groups., Adv. in Math. 11(1973), 289-303.

R. Bott, Some aspects of invariant theory in differential geometry. In: Differential Operators on Manifolds, C.I.M.E. 3 Ciclo 1975, 49-145.

R. Bott and A. Haefliger, On characteristic classes of Γ-foliations, Bull. Amer. Math. Soc. 78(1972), 1039-1044.

R. Bott and J. Heitsch, A remark on the integral cohomology of $B\Gamma_q$, Topology 11(1972), 141-146.

R. Bott, H. Shulman and J. Stasheff, On the de Rham theory of certain classifying spaces, Adv. Math. 20(1976), 43-56.

L. Bouma and W. van Est, Manifold schemes and foliations on the 2-torus and the Klein bottle, I, Proc. K. Ned. Akad. Wet., Ser. A 81(1978), 313-325.

L. Bouma and W. van Est, Manifold schemes and foliations on the 2-torus and the Klein bottle, II, Proc. K. Ned. Akad. Wet., Ser. A 81(1978), 326-338.

L. Bouma and W. van Est, Manifold schemes and foliations on the 2-torus and the Klein bottle, III, Proc. K. Ned. Akad. Wet., Ser. A 81(1978), 339-347.

R. Bowen, Unique ergodicity for foliations, Astérisque 40(1976), 11-16.

A. Brakhman, Foliations without limit cycles, Mat. Zametki 9(1971), 181-191.

D. Brill and F. Flaherty, Isolated maximal surfaces in space-time, Commun. Math. Phys. 50(1976), 157-165.

D. Brill and F. Flaherty, Maximizing properties of extremal surfaces in general relativity, Ann. Inst. Henri Poincaré, A, 28(1978), 335-347.

F. Brito, Une obstruction géométrique à l'existence de feuilletages de codimension 1 totalement géodésiques, J. Diff. Geom. 16(1981), 675-684.

F. Brito, A remark on minimal foliations of codimension two, Tôhoku Math. J. 36(1984), 341-350.

F. Brito, R. Langevin and H. Rosenberg, Intégrales de courbure sur une variété feuilletée, C. R. Acad. Sci. Paris 285(1977), 533-536.

F. Brito, R. Langevin and H. Rosenberg, Intégrales de courbure sur une variété feuilletée, J. Diff. Geom. 16(1981), 19-20.

F. Brito and P. Walczak, Totally geodesic foliations with integral normal bundles, preprint.

R. Brooks, Volumes and characteristic classes of foliations, Topology 18(1979), 295-304.

R. Brooks, Some Riemannian and dynamical invariants of foliations, Proc. of the 1981-82 year in Differential geometry, Univ. of Maryland, Birkhäuser, Progress in Math. 32(1983), 56-72.

R. Brooks, The spectral geometry of foliations, Amer. J. Math. 106(1984), 1001-1012.

R. Brooks and W. Goldman, The Godbillon-Vey invariant of a transversely homogeneous foliation, Trans. Amer. Math. Soc. 286(1984), 651-664.

H. Browne, Codimension 1 totally geodesic foliations of H^n, Tôhoku Math. J. 36(1984), 315-340.

J.-P. Buffet and J.-C. Lor, Une construction d'un universel pour une classe assez large de Γ-structures, C. R. Acad. Sci. Paris 270(1970), 640-642.

D. Burns, Curvatures of Monge-Ampère foliations, Ann. of Math. 115(1982), 349-373.

R. Caddeo, On the torsional cohomology of Molino for an almost complex manifold, Rend. Sem. Fac. Sci. Univ. Cagliari 50(1980), 765-777.

G. Cairns, Feuilletages riemanniens et classes caractéristiques fines et exotiques, Thèse 3^1 cycle, Montpellier (1982).

G. Cairns, Géométrie globale des feuilletages totalement géodésiques, C. R. Acad. Sci. Paris 297(1983), 525-527.

G. Cairns, Feuilletages totalement géodésiques de dimension 1, 2 ou 3, C. R. Acad. Sci. Paris 298(1984), 341-344.

G. Cairns, Feuilletages totalement géodésiques, Séminaire de géométrie différentielle 1983-84, Montpellier.

G. Cairns, Une remarque sur la cohomologie basique d'un feuilletage riemannien, Séminaire de géométrie différentielle 1984-85, Montpellier.

G. Cairns, Aspects cohomologiques des feuilletages totalement géodésiques, C. R. Acad. Sci. Paris 299(1984), 1017-1019.

G. Cairns, A general description of totally geodesic foliations, Tôhoku Math. J. 38(1986), 37-55.

G. Cairns, Feuilletages totalement géodésiques sur les variétés simplement connexes, preprint.

G. Cairns, Some properties of a cohomology group associated to a totally geodesic foliation, Math. Z. 192(1986), 391-403.

G. Cairns, Feuilletages géodésibles, Thèse, Univ. des Sciences et Techniques du Languedoc, Montpellier, 1987.

G. Cairns and E. Ghys, Totally geodesic foliations on 4-manifolds, J. Diff. Geom. 23(1986), 241-254.

B. Callenaere and D. Lehmann, Classes exotiques universelles, Ann. Inst. Fourier 24(1974), 301-306.

C. Camacho, Structural stability of foliations with singularities, Springer Lecture Notes in Math. 652(1978), 128-137.

C. Camacho, Singularities of holomorphic differential equations, Singularities and dynamical systems, Iraklion, 1983; North-Holland Math. Stud. 103(1985), 137-159.

C. Camacho and A. Neto, Geometric Theory of Foliations, Birkhäuser, Boston, 1985.

J. Cantwell and L. Conlon, Open leaves in closed 3-manifolds, Bull. Amer. Math. Soc. 82(1976), 256-258.

J. Cantwell and L. Conlon, Closed transversals and the genus of closed leaves in foliated 3-manifolds, J. Math. Anal. Appl. 55(1976), 653-657.

J. Cantwell and L. Conlon, Leaves with isolated ends in foliated 3-manifolds, Topology 16(1977), 311-322.

J. Cantwell and L. Conlon, Growth of leaves, Comment. Math. Helv. 53(1978), 93-111.

J. Cantwell and L. Conlon, Leaf prescriptions for closed 3-manifolds, Trans. Amer. Math. Soc. 236(1978), 239-261.

J. Cantwell and L. Conlon, Poincaré-Bendixson theory for leaves of codimension one, Trans. Amer. Soc. 265(1981), 181-209.

J. Cantwell and L. Conlon, Nonexponential leaves at finite level,
Trans. Amer. Math. Soc. 269(1982), 637-661.

J. Cantwell and L. Conlon, Smoothing fractional growth, Tôhoku Math.
J. 33(1981) 249-262.

J. Cantwell and L. Conlon, Analytic foliations and the theory of
levels, Math. Ann. 265(1983), 253-261.

J. Cantwell and L. Conlon, Foliations and subshifts, preprint.

J. Cantwell and L. Conlon, The dynamics of open, foliated manifolds
and a vanishing theorem for the Godbillon-Vey class, Advances in Math.
53(1984), 1-27.

J. Carballés, Characteristic homomorphism for (F_1,F_2)-foliated bundles
over subfoliated manifolds, Ann. Inst. Fourier 34(1984), 219-245.

A. Carfagna D'Andrea, A characterization of the tangent bundle of a
foliation, C. R. Acad. Sci. Paris 301(1985), 77-80.

M. Carfora, Zero-lapse loci in asymptotically flat maximally foliated
spacetime manifolds, Phys. Lett. A 84(1981), 53-55.

J. Carinena and L. Ibort, On Lax equations arising from Lagrangian
foliations, Lett. Math. Phys. 8(1984), 21-26.

P. Caron, Flots transversalement de Lie. Thèse de 3ème cycle,
Université de Lille (1980).

P. Caron et Y. Carrière, Flots transversalements de Lie $\mathbb{R}^n$, flots de
Lie minimaux, C. R. Acad. Sci. Paris 280(1980), 477-478.

F. Carreras, Linear invariants of Riemannian almost product manifolds,
Math. Proc. Camb. Phil. Soc. 91(1982), 99-106.

F. Carreras and A. Naveira, On the Pontrjagin algebra of a certain
class of flags of foliations, Canad. Math. Bull. 28(1985), 77-83.

Y. Carrière, Flots riemanniens et feuilletages géodésibles de
codimension un, Thèse Université des Sciences et Techniques de Lille I
(1981).

Y. Carrière, Flots riemanniens, Astérisque 116(1984), 31-52.

Y. Carrière, Les propriétés topologiques des flots riemanniens
retrouvées à l'aide du théorème des variétés presque plates, Math. Z.
186(1984), 393-400.

Y. Carrière, Sur la croissance des feuilletages de Lie, preprint.

Y. Carrière, Feuilletages riemanniens à croissance polynomiale,
preprint.

Y. Carrière et E. Ghys, Feuilletages totalement géodésiques, Anais da Acad. Bras. de Ciencias 53(1981), 427-432.

Y. Carrière et E. Ghys, Relations d'equivalence moyennables sur les groupes de Lie, C.R. Acad. Sci. Paris 300(1985), 677-680.

E. Cartan, Sur certaines expressions différentielles et le problème de Pfaff. Ann. Sci. École Norm. Sup. 16(1899). Oeuvres II, 303-396.

E. Cartan, Sur l'intégration des systèmes d'équations aux différentielles totales. Ann. Sci. École Norm. Sup. 18(1901), 241-311, Oeuvres II, 411-481.

H. Cartan, Cohomologie réelle d'un espace fibré princpal différentiable. In: Sém. Cartan 1949/50, exp. 19-20. Paris: Ecole Norm. Sup.

D. Cass, Minimal leaves in foliations, Trans. Amer. Math. Soc. 287(1985), 201-213.

I. Cattaneo-Gasparini, Global reduction of a dynamical system on a foliated manifold, J. Math. Phys. 25(1984), 2918-2921.

I. Cattaneo-Gasparini, Global reduction of a dynamical system on a foliated manifold and controlled projectability, Dynamical systems and microphysics, Academic Press, 1984, 183-205.

B. Cenkl, On the de Rham complex of $B\hat{\Gamma}$, Proc. Symp. Pure Math. Part 1, 27(1973), 265-274.

B. Cenkl, Residues of singularities of holomorphic foliations, J. Diff. Geom. 13(1978), 11-23.

B. Cenkl, Formulas for the characteristic classes of groups of diffeomorphisms, Rend. Mat.(7) 1(1981), 443-462.

D. Cerveau, Distributions involutives singuliéres, Ann. Inst. Fourier 29(1979), 261-294.

D. Cerveau, Integrating agents and the cobordism problem of germs of one-codimensional singular holomorphic foliations, Hokkaido Math. J. 14(1985), 21-32.

G. Chatelet, Sur les feuilletages induits par l'action de groupes de Lie nilpotents, Ann. Inst. Fourier 27(1977), 161-189.

G. Chatelet and H. Rosenberg, Un théorème de conjugaison des feuilletages, Ann. Inst. Fourier 21(1971), 95-106.

S. Cheng and S. Yau, Maximal spacelike hypersurfaces in the Lorentz-Minkowski space, Ann. of Math. 104(1976), 407-419.

B. Chen and P. Piccinni, The canonical foliations of a locally conformal Kähler manifold, Ann. Mat. Pura Appl. 141(1985), 289-305.

S. Chern, The geometry of G-Structures, Bull. Amer. Math. Soc. 72(1966), 167-219.

S. Chern and J. Simons, Characteristic forms and geometric invariants, Ann. of Math. 99(1974), 48-69.

S. Chern and K. Tenenblat, Foliations on a surface of constant curvature and the modified Korteweg-de Vries equation, J. Diff. Geom. 16(1981), 347-349.

W. Chow, Über Systeme von linearen partiellen Differentialgleichungen erster Ordnung. Math. Ann. 117(1940/41), 98-105.

D. Christodoulou and M. Francaviglia, Some dynamical properties of Einstein spacetimes admitting a Gaussian foliation, Gen. Relativity Gravitation 10(1979), 455-459.

P. Chrusciel, Sur les feuilletages conformément minimaux des variétés riemanniennes de dimension trois, C. R. Acad. Sci. Paris 301(1985), 609-612.

A. Clebsch, Uber die simultane Integration linearer partieller Differentialgleichungen, J. Reine Angew. Math. 65(1866), 257-268.

F. Cohen and L. Taylor, Computations of Gelfand-Fuks cohomology, the cohomology of function spaces, and the cohomology of configuration spaces, Springer Lecture Notes in Math. 657(1978), 106-143.

M. Cohen, Approximation of foliations, Can. Math. Bull. 14(1971), 311-314.

M. Cohen, Smoothing one-dimensional foliations on $S^2 \times S^2$, Can. Math. Bull. 16(1973), 43-44.

M. Cohen, Foliations on 3-manifolds, Amer. Math. Mon. 81(1974), 462-473.

L. Conlon, Transversally parallelizable foliations of codimension two, Trans. Amer. Math. Soc. 194 (1974), 79-102.

L. Conlon, Erratum to Transversally parallelizable foliations of codimension two, Trans. Amer. Math. Soc. 207(1975), 406.

L. Conlon, Foliations and exotic classes, Lectures at the Universidad de Extramadura, Jarandilla de la Vera (Caceres), 1985.

L. Conlon and S. Goodman, Opening closed leaves in foliated 3-manifolds, Topology 14(1975), 59-61.

L. Conlon and S. Goodman, The closed leaf index of foliated manifolds, Trans. Amer. Math. Soc. 233(1977), 205-223.

A. Connes, The von Neumann algebra of a foliation, Springer Lecture Notes in Phys. 80(1978), 145-151.

A. Connes, Sur la théorie noncommutative de l'intégration. In: algèbres d'opérateurs, pp. 19-143, Springer Lecture Notes in Math. 725(1979).

A. Connes, Feuilletages et algèbres d'opérateurs. In: Sém. Bourbaki 1979/80, exp. 551, Springer Lecture Notes in Math. 842(1980).

A. Connes, A survey of foliations and operator algebras. Proc. Symp. Pure Math 38, Amer. Math. Soc.,Part 1 (1982), 521-628.

A. Connes, Non commutative differential geometry, Publ. Math. IHES 62(1985), 41-144.

A. Connes and G. Skandalis, Théorème de l'indice pour les feuilletages, C. R. Acad. Sci. Paris 292(1981), 871-876.

L. Cordero, Nota sobre una decomposicion del operador diferencial exterior en una estructura casi-producto, Actas de la undécime reunion annal de matematicos Espanoles (1971), 124-129.

L. Cordero, Sur une théorie de cohomologie associée aux feuilletages, C. R. Acad. Sci. Paris 272(1971), 1056-1057.

L. Cordero, P-normal almost-product structure, Tensor 28(1974), 229-238.

L. Cordero, The extension of G-foliations to tangent bundles of higher order, Nagoya Math. J. 56(1975), 29-44.

L. Cordero, The horizontal lift of a foliation and its exotic classes, Springer Lecture Notes in Math. 484(1975), 192-200.

L. Cordero, Special connections on almost-multifoliate Riemannian manifolds, Math. Ann. 216(1975), 209-215.

L. Cordero, Sheaves and cohomologies associated to subfoliations, Resultate Math. 8(1985), 9-20.

L. Cordero (Ed), Differential Geometry, Proc. Colloq. Santiago de Compostela 1984, Pitman Research Notes 131(1985).

L. Cordero and X. Masa, Characteristic classes of subfoliations, Ann. Inst. Fourier 31(1981), 61-86.

L. Cordero and A. de Prada, Sur un feuillage dans le fibré tangent à une variété feuilletée, C. R. Acad. Sci. Paris 275(1972), 831-833.

L. Cordero and A. de Prada, Sobre las cohomologias y metricas del fibrado tangente a una variedad foliada, Actas Prim. J. Mat. Luso-Esp. Publs. Inst. 'Jorge Juan' Mat., Madrid (1973), 255-259.

L. Cordero and A. de Prada, Foliacion en el fibrado tangente a una variedad foliada y la obstruccion de Bott a la integrabilidad, Actas Prim. J. Mat. Luso-Es. Publs. Inst. 'Jorge Juan' Mat., Madrid (1973), 269-274.

L. Cordero and P. M. Gadea, Exotic characteristic classes and subfoliations, Ann. Inst. Fourier 26(1976), 225-237.

M. Craioveanu, Sur les sous-feuilletages d'une structure feuilletée, C. R. Acad. Sci. Paris 272(1971), 731-733.

M. Craioveanu, Variétés banachiques feuilletées, I, An. Univ. Timisoara, Ser. Sti. Mat. 9(1971), 35-48.

M. Craioveanu, Variétés banachiques feuilletées, II, An. Univ. Timisoara, Ser. Sti. Math. 13(1975), 11-12.

M. Craioveanu and M. Puta, Cohomology on a Riemannian foliated manifold with coefficients in the sheaf of germs of foliated currents, Math. Nachr. 99(1980), 43-53.

C. Curras-Bosch, Transformations in foliate manifolds, to appear.

C. Curras-Bosch, Infinitesimal transformations preserving a foliation, to appear.

C. Cumenge, Sheaves and cohomology of leaf spaces of foliations, C. R. Acad. Sci. Paris 297(1983), 195-198.

A. Davis and F. Wilson, Vector fields tangent to foliations, I, Reeb foliations, J. Diff. Eq. 11(1972), 491-498.

P. Dazord, Sur le géométrie des fibrés et des feuilletages lagrangiens, Ann. Sci. Ec. Norm. Sup. 14(1981), 465-480.

P. Dazord, Feuilletages et Mécanique Hamiltonienne, interventions de la géométrie en analyse et en physique mathématique, Université C. Bernard (1982).

P. Dazord, Sur l'existence de feuilles sphériques, C. R. Acad. Sci. Paris 296(1983), 77-79.

P. Dazord, Holonomie des feuilletages singuliers, C. R. Acad. Sci. Paris 298(1984), 27-30.

F. Deahna, Uber die Bedingungen der Integrabilität linearer Differentialgleichungen erster Ordnung zwischen einer beliebigen Anzahl Veränderlicher Grössen, J. reine angew. Math. 20(1840), 340-349.

K. Decesaro and T. Nagano, On compact foliations, Proc. Symp. Pure Math., Part 1, 27(1975), 277-281.

T. Delzant, Foliations of symplectic manifolds, C. R. Acad. Sci. Paris 300(1985), 201-204.

A. Denjoy, Sur les courbes défínis par les équations différentielles à la surface du tore, J. Math. Pures Appl. 11(1932), 333-375.

N. Desolneux-Moulis, Sur certaines familles à un paramètre de $T^2 \times S^2$, C. R. Acad. Sci. Paris 287(1978), 1043-1046.

P. Dippolito, Codimension one foliations of closed manifolds, Ann. of Math. 107(1978), 403-453.

P. Dombrowski, Jacobi fields, totally geodesic foliations and geodesic differential forms, Resultate der Math. 1(1978), 156-194.

P. Dombrowski, Classification up to diffeomorphism of measure preserving foliations of the torus $S^2 \times S^2$, following S. Sternberg, Singularities, foliations and Hamiltonian mechanics (Balaruc 1985), Travaux en cours, Hermann (1985), 1-19.

T. Duchamp, Characteristic invariants of G-foliations, Ph.D. Thesis, University of Illinois, Urbana, Illinois (1976).

T. Duchamp and M. Kalka, Holomorphic foliations and the Kobayashi metric, Proc. Amer. Math. Soc. 67(1977), 117-122.

T. Duchamp and M. Kalka, Holomorphic foliations and deformations of the Hopf foliation, Pac. J. of Math. 112(1984), 69-81.

T. Duchamp and M. Kalka, Invariants of tangentially holomorphic foliations and the Monge-Ampère equation, to appear.

J. Dufour (Ed), Singularités, feuilletages et mécanique hamiltonienne, Séminaire Sud-Rhodanien (1985).

G. Duminy, L'invariant de Godbillon-Vey d'un feuilletage se localise dans les feuilles ressort, preprint (1982).

G. Duminy and V. Sergiescu, Sur la nullité de l'invariant de Godbillon-Vey, C. R. Acad. Sci. Paris 292(1981), 821-824.

A. Durfee, Foliations of odd-dimensional spheres, Ann. of Math. 96(1972), 407-411.

A. Durfee and H. Lawson, Fibered knots and foliations of highly connected manifolds, Invent. Math. 17(1972), 203-215.

S. Duzhin, A spectral sequence connected with a foliation and cohomology of certain Lie algebras of vector fields, Uspekhi Mat. Nauk 39(1984), 135-136.

W. Dwyer, R. Ellis and R. Szczarba, Foliations with nonorientable leaves, Proc. Amer. Math. Soc. 89(1983), 733-738.

R. Edwards, A question concerning compact foliations, Springer Lecture Notes in Math. 468(1975), 2-4.

R. Edwards, K. Millett, and D. Sullivan, Foliations with all leaves compact, Topology 16(1977), 13-32.

C. Ehresmann, Les prolongements d'une variété différentiable. I. Calcul des jets. II. L'espace des jets d'ordre r de V_n dans V_m. III. Transitivité des prolongements, C. R. Acad. Sci. Paris 233(1951), 598-600, 777-779, 1081-1083.

C. Ehresmann, Sur la théore des variétés feuilletées, Rendiconti di Matematica e delle sue applicazioni, serie V. vol. X, 1-2, Roma, (1951).

C. Ehresmann, Structures locales et structures infinitésimales, C. R. Acad. Sci. Paris 243(1952), 587-589.

C. Ehresmann, Les prolongements d'une variété différentiable. IV. Eléments de contact et éléments d'envelope. V. Covariants différentiels et prolongements d'une structure infinitésimale, C. R. Acad. Sci. Paris 234(1952), 1028-1030, 1424-1425.

C. Ehresmann, Structures feuilletées, Proc. Fifth Canad. Math. Congress (1961).

C. Ehresmann and G. Reeb, Sur les champs d'éléments de contact de dimension p complètement intégrables dans une variété continuement différentiable, C. R. Acad. Sci. Paris 218(1944), 955-957.

C. Ehresmann and Shi-Weishu, Sur les espaces feuilletées: théorème de stabilité, C. R. Acad. Sci. Paris 243(1956), 344-346.

J. Eells, Elliptic operators on manifolds, complex analysis and its appl. (Trieste), I(1975), 95-152.

J. Eells and L. Lemaire, A report on harmonic maps, Bull. London Math. Soc. 10(1978), 1-68.

J. Eells and J. Sampson, Harmonic mappings of Riemannian manifolds, Amer. J. Math. 86(1964), 109-160.

A. E. Kacimi-Alaoui, Sur la cohomologie feuilletée, Comp. Math. 49(1983), 195-215.

A. El Kacimi-Alaoui and G. Hector, Décomposition de Hodge sur l'espace des feuilles d'un feuilletage riemannien, C. R. Acad. Sci. Paris 298(1984), 289-292.

A. El Kacimi-Alaoui and G. Hector, Décomposition de Hodge basique pour un feuilletage riemannien, Ann. Inst. Fourier 36(1986), 207-227.

A. El Kacimi-Alaoui, V. Sergiescu et G. Hector, La cohomologie basique d'un feuilletage riemannien est de dimension finie, Math. Z. 188(1985), 593-599.

C. Ennis, Sufficient conditions for smoothing codimension one foliations, Trans. Amer. Math. Soc. 276(1983), 311-322.

C. Ennis, M. Hirsch and C. Pugh, Foliations that are not approximable by smoother ones, Springer Lecture Notes in Math. 1007(1983), 146-176.

C. Epstein, Orthogonally integrable line fields in H^3, Comm. Pure Appl. Math. 38(1985), 593-608.

D. Epstein, The simplicity of certain groups of homeomorphisms, Compos. Math. 22(1970), 165-173.

D. Epstein, Periodic flows on three-dimensional manifolds, Ann. of Math. 95(1972), 66-82.

D. Epstein, Foliations with all leaves compact, Springer Lecture Notes in Math. 468(1975), 1-2.

D. Epstein, Foliations with all leaves compact, Ann. Inst. Fourier 26(1976), 265-282.

D. Epstein, A topology for the space of foliations, Springer Lecture Notes in Math. 597(1977), 132-150.

D. Epstein, Transversely hyperbolic 1-dimensional foliations, Astérisque 116(1984), 53-69.

D. Epstein, K. Millet and D. Tischler, Leaves without holonomy, J. London Math. Soc. 16(1977), 548-552.

D. Epstein and H. Rosenberg, Stability of compact foliations, Springer Lecture Notes in Math. 597(1978), 151-160.

D. Epstein and E. Vogt, A counterexample to the periodic orbit conjecture in codimension 3, Ann. Math. 108(1978), 539-552.

R. Escobales, Riemannian submersions with totally geodesic fibers, J. Diff. Geom. 10(1975), 253-276.

R. Escobales, Sufficient conditions for a bundle-like foliation to admit a Riemannian submersion onto its leaf space, Proc. Amer. Math. Soc. 84(1982), 280-284.

R. Escobales, The integrability tensor for bundle-like foliations, Trans. Amer. Math. Soc. 270(1982), 333-339.

R. Escobales, Bundle-like foliations with Kählerian leaves, Trans. Amer. Math. Soc. 276(1983), 853-859.

R. Escobales, Riemannian foliations of the rank one symmetric spaces, Proc. Amer. Math. Soc. 95(1985), 495-498.

R. Escobales and P. Parker, Geometric consequences of the normal curvature cohomology class in umbilic foliations, to appear.

W. van Est, Group cohomology and Lie algebra cohomology in Lie groups. Nederl. Akad. Wetensch. Indag. Math. 15(1953), 484-492, 493-504.

W. van Est. On the algebraic cohomology concepts in Lie groups. Nederl. Akad. Wetensch. Indag. Math. 17(1955), 225-233, 286-294.

W. van Est, Une application d'une méthode de Cartan-Leray, Nederl. Akad. Wetensch. Indag. Math. 17(1955), 542-544.

W. van Est, A generalization of the Cartan-Leray spectral sequence, Nederl. Akad. Wetensch. Indag. Math. 20(1958), 399-413.

W. van Est, Fundamental group of manifold schemes, Topological structures, II, Part 1, Math. Centre Tracts Amsterdam 115(1979), 79-90.

H. Farran, G-structures on manifolds with parallel foliations, J. Univ. Kuwait Sci. 7(1980), 59-67.

H. Farran, On Anosov foliations, Math. Rep. Toyama Univ. 7(1984), 1-11.

L. Favaro, Differentiable mappings between foliated manifolds, Bol. Soc. Brasil. Mat. 8(1977), 39-46.

E. Fedida, Sur les feuilletages de Lie, C. R. Acad. Sci. Paris 272(1971), 999-1002.

E. Fedida, Structures différentiables sur le branchement simple et équations différentielles dans le plan, C. R. Acad. Sci. Paris 276(1973), 1657-1659.

E. Fedida, Sur l'existence des feuilletages de Lie, C. R. Acad. Sci. Paris 278(1974), 835-837.

E. Fedida, Sur la théorie des feuilletages associées au repère mobile: cas des feuilletages de Lie, Springer Lecture Notes in Math. 652(1978), 183-195.

E. Fedida, and P. Furness, Feuilletages transversalement affine de codimension 1, C. R. Acad. Sci. Paris 282(1976), 825-827.

E. Fedida and P. Furness, Transversally affine foliations, Glasgow Math. J. 17(1976), 106-111.

E. Fedida, C. Hyjazi and F. Pluvinage, Sur les feuilletages transverses du plan, Afrika Mat. 7(1985), 63-87.

B. Feigin, Characteristic classes of flags of foliations, Funkts. Anal. Ego Prilozhen. 9(1975), 49-56.

B. Feigin and D. Fuks, Stable cohomology of the algebra W_n and relations in the algebra L_1, Funktsional. Anal. i Prilozen 18(1984), 94-95.

D. Ferus, Totally geodesic foliations, Math. Ann. 188(1970), 313-316.

D. Ferus, On the completeness of nullity foliations, Mich. Math. J. 18(1971), 61-64.

M. Fliess, Cascade decomposition of nonlinear systems, foliations and ideals of transitive Lie algebras, Systems Control Lett. 5(1985), 263-265.

J. Franks, Two foliations in the plane, Proc. Amer. Math. Soc. 58(1976) 262-264.

J. Franks, Holonomy invariant cochains for foliations, Proc. Amer. Math. Soc. 62(1977), 161-164.

M. Freeman, Local complex foliation of real submanifolds, Math. Ann. 209(1974) 1-30.

M. Freeman, The Levi form and local complex foliations, Proc. Amer. Math. Soc. 57(1976), 369-370.

F. Frobenius, Über das Pfaffsche Problem, J. reine Angew. Math. 82(1877), 267-282. Ges. Abh. I, 286-301.

D. Fuks, Characteristic classes of foliations, Usp. Mat. Nauk, 28(1973), 3-17.

D. Fuks, Finite-dimensional Lie algebras of formal vector fields and characteristic classes of homogeneous foliations, Izv. Akad. Nauk SSSR, Ser. Mat. (1976), 40.

D. Fuks, Cohomology of infinite-dimensional Lie algebras and characteristic classes of foliations, Itogi Nauki-Seriya "Matematika" 10(1976), 179-286 [Russian]. Translation: J. Soviet Math. 11(1979), 922-980.

D. Fuks, A. Gabrielov and I. Gel'fand, The Gauss-Bonnet theorem and the Atiyah-Patodi-Singer functionals for the characteristic classes of foliations, Topology 15(1976), 165-188.

D. Fuks, Non-trivialité des classes caractéristiques des g-structures, Applications aux classes caractéristiques des feuilletages, C. R. Acad. Sci. Paris 284(1977), 1017-1019.

D. Fuks, Non-trivialité des classes caractéristiques des g-structures, Applications aux variations des classes caractéristiques de feuilletages, C. R. Acad. Sci. Paris 284(1977), 1105-1107.

D. Fuks, Foliations, Itogi Nauki-Seriya Algebra, Topologiya, Geometriya 18(1981), 151-213 [Russian]. Translation: J. Soviet Math. 18(1982), 255-291.

K. Fukui, Codimension 1 foliations on simply connected 5-manifolds, Proc. Japan Acad. 49(1973), 432-434.

K. Fukui, An application of the Morse theory to foliated manifolds, Nagoya Math. J. 54(1974), 165-178.

K. Fukui, On the foliated cobordisms of codimension-one foliated 3-manifolds, Acta Hum. Sci. Univ., Sangio Kiotiensis Nat. Sci. Ser. 7(1978), 42-49.

K. Fukin, A remark on the foliated cobordism of codimension-one foliated 3-manifolds, J. Math. Kyoto Univ. 18(1978), 189-197.

K. Fukui, Perturbations of compact foliations, Proc. Japan Acad. Ser. A. 58(1982), 341-344.

K. Fukui, Stability and instability of certain foliations of 4-manifolds by closed orientable surfaces, Publ. RIMS, Kyoto Univ. 22(1986), 1155-1171.

K. Fukui, Stability of foliations of 3-manifolds by circles, J. Math. Soc. Japan 39(1987), 117-126.

K. Fukui and N. Tomita, Lie algebra of foliation preserving vectorfields, J. Math. Kyoto Univ. 22(1983), 685-699.

P. Furness, Affine foliations of codimension one, Q. J. Math. No. 98, 25(1974), 151-161.

A. Futaki and S. Morita, Invariant polynomials on compact complex manifolds, Proc. Japan Acad. Ser. A Math. Sci. 60(1984), 369-372.

D. Gabai, Foliations and the topology of 3-manifolds, Bull. Amer. Math. Soc. 8(1983), 77-80.

D. Gabai, Foliations and the topology of 3-manifolds, J. Diff. Geom. 18(1983), 445-503.

D. Gabai, Foliations and the genera of links, Topology 23(1984), 381-394.

M. Garançon, Homotopie et holonomie de certains feuilletages de codimension 1, Ann. Inst. Fourier 22(1972), 61-71.

M. Garançon, Feuilletages transversalement analytiques de codimension 1 admettant une transversale fermée qui coupe toutes les feuilles, Ann. Inst. Fourier 22(1972), 271-287.

R. Gardner, Differential geometry and foliations: the Godbillon-Vey invariant and the Bott-Pasternack vanishing theorems, Springer Lecture Notes in Math. 652(1978), 75-94.

L. Garnett, Statistical properties of foliations, Springer Lecture Notes in Math. 1007(1983), 294-299.

L. Garnett, Functions and measures harmonic along the leaves of a foliation and the ergodic theorem, J. Funct. Analys. 51(1983), 285-311.

D. Gauld, Foliations on topological manifolds, Math. Chron. 2(1972), 29-41.

I. Gelfand and D. Fuks, Cohomologies of the Lie algebra of tangent vector fields of a smooth manifold, Funkts. Anal. i Ego Prilozen. 3(1969), 32-52.

I. Gelfand and D. Fuks, The Cohomologies of the Lie algebra of formal vector fields, Izv. Akad. Nauk SSSR, Ser. Mat. 34(1970), 322-337.

I. Gelfand, D. Kalinin, and D. Fuks, On the cohomology groups of the Lie algebra of Hamiltonian formal vector fields. Funkcional. Anal. i Ego Prilozen 6, 25-29[Russian]. Translation: Functional Anal. Apply. 6(1972), 193-196.

I. Gelfand, D. Kazhdan, and D. Fuks, The actions of infinite dimensional Lie algebras. Funkcional. Anal. i Ego Prilozen 6, 10-15 [Russian], Functional Anal. Appl. 6(1972), 9-13.

I. Gelfand and D. Fuks, PL-foliations, Funkcional. Anal. i Ego Prilozen. 7(1973), 29-37.

I. Gelfand and D. Fuks, PL-foliations, II, Funkcional. Anal. i Ego Prilozen 8(1974), 7-11.

I. Gelfand, B. Feigin and D. Fuks, Cohomologies of the Lie algebra of formal vector fields with coefficients in the dual space and variations of characteristic classes of foliation, Funkcional. Anal. i Ego Prilozen 8(1974), 13-29.

C. Gerhardt, H-surfaces in Lorentzian manifolds, Commun. Math. Phys. 89(1983), 523-553.

E. Ghys, Classification des feuilletages totalement géodésiques de codimension un, Comment. Math. Helv. 58(1983), 543-572.

E. Ghys, Feuilletages riemanniens sur les variétés simplement connexes, Ann. Inst. Fourier 34(1984), 203-223.

E. Ghys, Groupes d'holonomie des feuilletages de Lie, Proc. Kon. Nederl. Acad. Sci. A 88(1985), 173-182.

E. Ghys, Un feuilletage analytique dont la cohomologie basique est de dimension infinie, preprint.

E. Ghys and V. Sergiescu, Stabilité et conjugaison différentiable pour certains feuilletages, Topology 19(1980), 179-197.

E. Ghys and V. Sergiescu, Sur un groupe remarquable de difféomorphismes du cercle, IHES preprint (1986).

0. Gil-Medrano and A. Naveira, The Gauss-Bonnet integrand for a class of Riemannian manifolds admitting two orthogonal complementary foliations, Canad. Math. Bull. 26(1983), 358-364.

O. Gil-Medrano and A. Naveira, Some remarks about the Riemannian curvature operator of a Riemannian almost-product manifold, Rev. Roumaine Math. Pures Appl. 30(1985), 647-658.

J. Girbau, Vanishing theorems for complex analytic foliate manifolds, Proc. IV. Int. Coll. Diff. Geom., Santiago de Compostela 1978, 131-140.

J. Girbau, Vanishing theorems and stability of complex analytic foliations, Springer Lecture Notes in Math. 792(1980), 247-251.

J. Girbau, Sur le théorème de stabilité de feuilletages de Hamilton, Epstein et Rosenberg, C. R. Acad. Sci. Paris 291(1980), 41-44.

J. Girbau, Vanishing cohomology theorems and stability of complex analytic foliations, Israel J. Math. 40(1981), 235-254.

J. Girbau, Some examples of deformations of transversely holomorphic foliations, Springer Lecture Notes in Math. 1045(1984), 53-62.

J. Girbau, A. Haefliger and A. Sundararaman, On deformations of transversely holomorphic foliations, Crelle J. 345(1983), 122-147.

J. Girbau and M. Nicolau, Pseudodifferential operators on V-manifolds and foliations, I, Collect. Math. 30(1979), 247-265.

J. Girbau and M. Nicolau, Pseudodifferential operators on V-manifolds and foliations, II, Collect. Math. 31(1980), 63-95.

J. Girbau and M. Nicolau, Deformations of holomorphic foliations and transvcrscly holomorphic foliations, Pitman Research Notes in Math. 131(1985), 162-173.

M. Giry, Study of nontransversally complete foliations in a regular case, Bull. Sci. Math. 106(1982), 433-442.

H. Gluck, Can space be filled by geodesics, and if so, how?, open letter (1979).

H. Gluck, Dynamical behavior of geodesic fields, Springer Lecture Notes in Math. 819(1980), 190-215.

H. Gluck and F. Warner, Great circle fibrations of the three sphere, Duke Math. J. 50(1983).

C. Godbillon, Feuilletages ayant la propriété du prolongement des homotopies, Thèse Doct. Sc. Math., Fac. Sci. Univ. Strasbourg (1967).

C. Godbillon, Problèmes d'existence et d'homotopie dans les feuilletages, Springer Lecture Notes in Math. 244(1971), 167-181.

C. Godbillon, Cohomologies d'algèbres de Lie de champs de vecteurs formels. In: Sém. Bourbaki 1972/73 exp. 421. Springer Lecture Notes in Math. 383(1972).

C. Godbillon, Fibrés en droites et feuilletages du plan, Enseign. Math. 18(1972-73), 213-224.

C. Godbillon, Cohomologies d'algèbres de Lie de champs de vecteurs formels, Springer Lecture Notes in Math. 383(1974), 69-87.

C. Godbillon, Feuilletages de Lie, Springer Lecture Notes in Math. 392(1974), 10-13.

C. Godbillon, Invariants of foliations, Global Anal. Appl. Lect. Int. Semin. Course Trieste, Vienna 2(1974), 215-219.

C. Godbillon and J. Vey, Un invariant des feuilletages de codimension un, C. R. Acad. Sci. Paris 273(1971), 92-95.

A. Goddard, Foliations of space-time by space-like hypersurfaces of constant mean curvature, Commnn. Math. Phys. 54(1977), 279-282.

B. Golbus, On the singularities of foliations and of vector bundle maps, Bol. Soc. Brasil. Mat. 7(1976), 11-35.

B. Golbus, On extending local foliations, Q. J. Math., No. 110, 28(1977), 163-176.

R. Goldman, The holonomy ring on the leaves of foliated manifolds, J. Diff. Geom. 11(1976), 411-449.

W. Goldmann, M. Hirsch and G. Levitt, Invariant measures for affine foliations, Proc. Amer. Math. Soc. 86(1982), 511-518.

X. Gómez-Mont, Transversal holomorphic structures, J. Diff. Geom. 15(1980), 161-186.

X. Gómez-Mont, On families of rational vector fields, Coll. on Dyn. Systems, Guanajuato, 1983, Aportaciones Mat., Soc. Mat. Mexicana 1985, 36-65.

S. Goodman, Closed leaves in foliated 3-manifolds, Proc. Nat. Acad. Sci. 71(1974), 4414-4415.

S. Goodman, Closed leaves in foliations of codimension one, Comment. Math. Helv. 50(1975), 383-388.

S. Goodman, On the structure of foliated 3-manifolds separated by a compact leaf, Invent. Math. 39(1977), 213-221.

S. Goodman and J. Plante, Holonomy and averaging in foliated sets, J. Diff. Geom. 14(1979), 401-407.

V. Gordin, Complex foliations of submanifolds, Usp. Mat. Nauk. 27(1972), 233-234.

R. Grimaldi, The asymptotic geometry of the leaves of a foliation, Rend. Circ. Mat. Palermo 32(1983), 199-207.

R. Grimaldi and G. Passante, The asymptotic geometry of the leaves of a foliation, Atti Acad. Sci. Torino Cl. Sci. Fis. Mat. Natur. 118(1984), 97-100.

R. Grimaldi and G. Passante, Asymptotic geometry for Anosov foliations, C. R. Acad. Sci. Paris 300(1985), 275-276.

D. Gromoll and K. Grove, One dimensional metric foliations in constant curvature spaces, Diff. Geom. and Complex Analysis, Rauch Memorial Volume 1985, 165-168.

M. Gromov, Transversal mappings of foliations, Dokl. Akad. Nauk SSSR 182, 255-258[Russian]. Translation: Soviet Math. Dokl. 9(1968), 1126-1129.

M. Gromov, Stable mappings of foliations into manifolds, Izv. Akad. Nauk SSSR, Ser. Mat. 33(1969), 707-734.

S. Guelorget, Algèbre de Weil du groupe linéaire, Application aux classes caractéristiques d'un feuilletage, Springer Lecture Notes in Math. 484(1975), 179-191.

S. Guelorget and G. Joubert, Algèbre de Weil et classes caractéristiques d'un feuilletage, C. R. Acad. Sci. Paris 277(1973), 11-14.

V. Guillemin, Remarks on some results of Gelfand and Fuks, Bull. Amer. Math. Soc. 78(1972), 539-540.

V. Guillemin, Cohomology of vector fields on a manifold, Adv. in Math. 10(1973), 192-220.

A. Haefliger, Sur les feuilletages analytiques, C. R. Acad. Sci. Paris 242(1956), 2908-2910.

A. Haefliger, Structures feuilletées et cohomologie à valeurs dans un faisceau de groupoides, Comment. Math. Helv. 32(1958), 249-329.

A. Haefliger, Variétés feuilletées, Ann. Scuola Norm. Sup. Pisa 16(1962), 367-397.

A. Haefliger, Travaux de Novikov sur les feuilletages. In: Sém. Bourbaki 1967/68, exp. 339, New York: W. A. Benjamin, (1968).

A. Haefliger, Feuilletages sur les variétés ouvertes, Topology 9(1970), 183-194.

A. Haefliger, Homotopy and integrability, Springer Lecture Notes in Math. 197(1971), 133-163.

A. Haefliger, Lectures on Gromov's theorem, Liverpool Singularities Symposium, Springer Lecture Notes in Math. 209(1971), 128-141.

A. Haefliger, Teorema de Bott sobre una obstruccion topologica a la integrabilidad completa, Rev. Mat. Hisp.-Amer. 32(1972), 21-32.

A. Haefliger, Sur les classes caractéristiques des feuilletages, Sém. Bourbaki 412-01 to 412-21(1972).

A. Haefliger, Sur les classes caractéristiques des feuilletages, Springer Lecture Notes in Math. 317(1973), 239-260.

A. Haefliger, Differentiable cohomology, C.I.M.E. Lectures 1976.

A. Haefliger, Sur la cohomologie de Lie des champs de vecteurs. Ann. Sci. Ecole Norm. Sup. 9(1976), 503-532.

A. Haefliger, Feuilletages avec feuilles minimales, Proceedings of the IV International Colloqium on Differential Geometry, University of Santiago de Compostela, 1979, 275-284.

A. Haefliger, Whitehead products and differential forms, Springer Lecture Notes in Math. 652(1978), 13-24.

A. Haefliger, On the Gelfand-Fuks cohomology, Enseign. Math. 24(1978), 143-160.

A. Haefliger, Cohomology of Lie algebras and foliations, Springer Lecture Notes in Math. 652(1978), 1-12.

A. Haefliger, Some remarks on foliations with minimal leaves, J. Diff. Geom. 15(1980), 269-284.

A. Haefliger, Groupoïdes d'holonomie et classifiants, Astérisque 116(1984), 70-97.

A. Haefliger, Pseudogroup of local isometries, Pitman Research Notes 131(1985), 174-197.

A. Haefliger and L. Banghe, Currents on a circle invariant by a Fuchsian group, Springer Lecture Notes in Math. 1007(1983), 369-378.

A. Haefliger and K. Sithanantham, A proof that $B\Gamma_1^c$ is 2-connected, Amer. Math. Soc., Contemp. Math. 12(1982), 129-139.

A. Haefliger and D. Sundararaman, Complexifications of transversely holomorphic foliations, Math. Ann. 272(1985), 23-27.

A. Hamasaki, Continuous cohomologies of Lie algebras of formal G-invariant vector fields and obstructions to lifting foliations, Publ. RIMS, Kyoto Univ. 20(1984), 401-429.

R. Hamilton, Deformation theory of foliations, preprint Cornell University (1978).

Th. Hangan and R. Lutz, Champs d'hyperplans totalement géodésiques sur les sphères, III^1 rencontre de géométrie du Schnepfenried, 1982, Astérisque 107-108(1983), 189-200.

D. Hardorp, All compact orientable three dimensional manifolds admit total foliations, Memoirs Amer. Math. Soc. 233(1980).

J. Harrison, Opening closed leaves of foliations, Bull. London Math. Soc. 15(1983), 218-220.

R. Harvey and H. Lawson, Calibrated foliations, Amer. J. of Math. 104(1982), 607-633.

J. Hass, Minimal surfaces in foliated manifolds, Comment. Math. Helv. 61(1986), 1-32.

J. Hausmann, Extension d'une homotopie à un feuilletage topologique, Manifolds-Amsterdam 1970, Springer Lecture Notes in Math 197(1971), 164-175.

J. Hebda, Curvature and focal points in Riemannian foliations, Indiana Univ. Math. J. 35(1986), 321-331.

G. Hector, Ouverts incompressibles et théorème de Denjoy-Poincaré pour les feuilletages, C. R. Acad. Sci. Paris 274(1972), 159-162.

G. Hector, Ouverts incompressibles et théorème de Denjoy-Poincaré pour les feuilletages, Thesis, University of Strasbourg 1972.

G. Hector, Ouverts incompressibles et structure des feuilletages de codimension 1, C. R. Acad. Sci. 274(1972), 741-744.

G. Hector, Sur les feuilletages presque sans holonomie, C. R. Acad. Sci. Paris 274(1972), 1703-1706.

G. Hector, Actions de groupes de difféomorphismes de $\{0,1\}$, Springer Lecture Notes in Math. 392(1974), 14-22.

G. Hector, Germes de feuilletages, Proc. of the 10th Braz. Math. Colloq. 1975, Vol. II, Inst. Mat. Pura Apl. Pocos de Caldas 1978, 547-552.

G. Hector, Quelques exemples de feuilletages d' espèces rares, Ann. Inst. Fourier 26(1976), 239-264.

G. Hector, Feuilletages en cyclindres, Springer Lecture Notes in Math. 597(1977), 252-270.

G. Hector, Leaves whose growth is neither exponential nor polynomial, Topology 16(1977), 451-459.

G. Hector, Feuilletages en cylindres. In: Geometry and Topology, Proc. III, Latin Amer. School of Math., Springer Lecture Notes in Math. 597(1977), 252-270.

G. Hector, Cohomology of Lie algebras and foliations, Springer Lecture Notes in Math. 652(1978), 1-12.

G. Hector, Croissance des feuilletages presque sans holonomie, Springer Lecture Notes in Math. 652(1978), 141-182.

G. Hector, Architecture of C^2 foliations, Astérisque 107-108 (1983), 243-258.

G. Hector and W. Bouma, All open surfaces are leaves of simple foliations of $\mathbb{R}^3$, Nederl. Akad. Wetensch. Indag. Math. 45(1983), 443-452.

G. Hector and U. Hirsch, Introduction to the Geometry of Foliations, Part A and B, Vieweg, Braunschweig, 1981 and 1983.

J. Heitsch, The cohomologies of classifying spaces for foliations, Thesis, University of Chicago, Chicago, IL (1971).

J. Heitsch, Deformations of secondary characteristic classes, Topology 12(1973), 381-388.

J. Heitsch, A cohomology for foliated manifolds, Bull. Amer. Math. Soc. 79(1973), 1283-1285.

J. Heitsch, A cohomology for foliated manifolds, Comment. Math. Helv. 50(1975), 197-218.

J. Heitsch, A remark on the residue theorem of Bott, Indiana Univ. Math. J. 25(1976), 1139-1147.

J. Heitsch, Residues and characteristic classes of foliations, Bull. Amer. Math. Soc. 83(1977), 397-399.

J. Heitsch, Independent variations of secondary classes, Ann. of Math. 108(1978), 421-460.

J. Heitsch, Derivatives of secondary characteristic classes, J. Diff. Geom. 13(1978), 311-339.

J. Heitsch, A residue formula for holomorphic foliations, Mich. Math. J. 27(1980), 181-194.

J. Heitsch, Linearity and residues for foliations, Bol. Soc. Bras. Mat. 12(1981), 87-94.

J. Heitsch, Flat bundles and residues for foliations, Invent. Math. 73(1983), 271-285.

J. Heitsch, Secondary invariants of transversely homogeneous foliations, Mich. Math. J. 33(1986) 47-57.

J. Heitsch and S. Hurder, Secondary classes, Weil measures and the geometry of foliations, J. Diff. Geom. 20(1984), 291-309.

D. Henc, Transverse intersections and stability of foliations, Bol. Soc. Brasil Mat. 13(1982), 1-18.

M. Hermann, Conjugaison C^∞ des difféomorphismes du cercle dont le nombre de rotation satisfait à une condition arithmétique, C. R. Acad. Sci. Paris 282(1976), 503-506.

M. Hermann, Conjugasion C^∞ des difféomorphismes du cercle pour presque tout nombre de rotation, C. R. Acad. Sci. Paris (1976), 579-582.

M. Hermann, The Godbillon-Vey invariant of foliations by planes of T^3, Springer Lecture Notes in Math. 597(1977), 294-307.

M. Hermann, Sur la conjugasion différentiable des difféomorphismes du cerle à des rotations, Inst. Hautes Etudes Sci. Publ. Math. 49(1979), 5-233.

R. Hermann, A sufficient condition that a mapping of Riemannian manifolds be a fiber bundle. Proc. Amer. Math. Soc. 11(1960), 236-242.

R. Hermann, The differential geometry of foliations I, Ann. of Math. 72(1960), 445-457.

R. Hermann, The differential geometry of foliations II, J. Math. Mech. 11(1962), 305-315.

R. Hermann, Totally geodesic orbits of groups of isometries, Indag. Math. 24(1962), 291-298.

R. Hermann, The Born theory of rigid motions in relativity and the theory of Riemannian foliations. In: Development of Mathematics in the Nineteenth Century, Appendices, Kleinian Mathematics from an Advanced Standpoint, 549-573. Brookline, MA: Math Sci. (1979).

S. Hiepko and H. Reckziegel, Über sphärische Blätterungen und die Vollständigkeit ihrer Blätter, Manuscripta Math. 31(1980), 269-283.

M. Hirsch, Stability of compact leaves of foliations, Proc. Internat. Conf. on Dynamical Systems, Salvador, Brazil, 1971, Academic Press, 135-155.

M. Hirsch, Foliated bundles, flat manifolds and invariant measures, Springer Lecture Notes in Math. 468(1975), 8-9.

M. Hirsch, A stable analytic foliation with only exceptional minimal sets, Springer Lecture Notes in Math. 468(1975), 9-10.

M. Hirsch and C. Pugh, Smoothness of horocycle foliations, J. Diff. Geom. 10(1975), 225-238.

M. Hirsch and W. P. Thurston, Foliated bundles, invariant measures and flat manifolds, Ann. of Math. 101(1975), 369-390.

U. Hirsch, Some remarks on analytic foliations and analytic branched coverings, Math. Ann. 248(1980), 139-152.

H. Holmann, Holomorphe Blätterungen komplexer Räume, Comment. Math. Helv. 47(1972), 185-204.

H. Holmann, On the stability of holomorphic foliations with all leaves compact, Springer Lecture Notes in Math. 683(1978) 217-248.

H. Holmann, Holomorphic transformation groups with compact orbits, Springer Lecture Notes in Math. 743(1979), 419-430.

H. Holmann, On the stability of holomorphic foliations, Springer Lecture Notes in Math. 798(1980), 192-202.

H. Holmann, Feuilletages complexes et symplectiques, Riv. Mat. Univ. Parma 10(1984), 91-108.

L. Hörmander, The Frobenius-Nirenberg theorem, Ark. Mat. 5(1964), 425-432.

S. Hurder, On the homotopy and cohomology of the classifying space of Riemannian foliations, Proc. Amer. Math. Soc. 81(1981), 484-489.

S. Hurder, Dual homotopy invariants of G-foliations, Topology 20(1981), 365-387.

S. Hurder, On the secondary classes of foliations with trivial normal bundles, Comment. Math. Helv. 56(1981), 307-326.

S. Hurder, Independent rigid secondary classes for holomorphic foliations, Invent. Math. 66(1982), 313-323.

S. Hurder, Vanishing of secondary classes for compact foliations, J. London Math. Soc. 28(1983). 175-183.

S. Hurder, The classifying space of smooth foliations, ILL. J. of Math. 29(1985), 108-133.

S. Hurder, Growth of leaves and secondary invariants of foliations, preprint.

S. Hurder, Exotic classes for measured foliations, Bull. Amer. Math. Soc. 7(1982), 389-391.

S. Hurder, Global invariant for measured foliations, Trans. Amer. Math. Soc. 280(1983), 367-391.

S. Hurder, The Godbillon measure for amenable foliations, preprint.

S. Hurder and F. Kamber, Homotopy invariants of foliations, Topology Siegen 1979, Springer Lecture Notes in Math. 788(1980), 49-61.

S. Hurder and A. Katok, Secondary classes and transverse measure theory of a foliation, Bull. Amer. Math. Soc. 11(1984), 347-350.

S. Hurder and A. Katok, Smoothness and Godbillon-Vey classes of geodesic horocycle foliation on surfaces of variable negative curvature, to appear.

M. Hvidsten and Ph. Tondeur, A characterization of harmonic foliations by variations of the metric, Proc. Amer. Math. Soc. 98(1986), 359-362.

V. Igosin, Decomposition theorems for double foliations that are compatible with pulverizations, Math. Notes 28(1980), 916-918.

V. Igosin and J. Sapiro, Stability of fibres of a foliation with a compatible Riemannian metric, Mat. Zametki 27(1980), 767-778, 830.

G. Ikegami, Existence of regular coverings associated with leaves of codimension one foliations, Nagoya Math. J. 67(1977), 15-34.

H. Imanishi, Sur l'existence des feuilletages S^2-invariants, J. Math. Kyoto Univ. 12(1972), 297-305.

H. Imanishi, On the theorem of Denjoy-Sacksteder for codimension one foliations without holonomy, J. Math. Kyoto Univ. 14(1974), 607-634.

H. Imanishi, On codimension one foliations defined by closed one-forms with singularities, J. Math. Kyoto Univ. 19(1979), 285-291.

H. Imanishi, Structure of codimension 1 foliations without holonomy on manifolds with abelian fundamental group, J. Math. Kyoto Univ. 19(1979), 481-495.

T. Inaba, On stability of proper leaves of codimension one foliations, J. Math. Soc. Japan 29(1977), 771-778.

T. Inaba, On the structure of real analytic foliations of codimension one, J. Fac. Sci. Univ. Tokyo 26(1979), 453-464.

T. Inaba, C^2 Reeb stability of non compact leaves of foliations, Proc. Japan Acad. 59(1983), 158-160.

T. Inaba, Reeb stability for noncompact leaves, Topology 22(1983), 105-118.

T. Inaba, T. Nishimori, M. Takamura and N. Tsuchiya, Open manifolds which are nonrelizable as leaves, Kodai Math. J. 8(1985), 112-119.

J. Isenberg and V. Moncrief, The existence of constant mean curvature foliations of Gowdy 3-torus spacetime, Comm. Math. Phys. 86(1982), 485-493.

D. Ivanenko and G. Sardanashvily, Foliation analysis of gravitation singularities, Phys. Lett. A 91(1982), 341-344.

S. Izumiya, Smooth mapings between foliated manifolds, Bol. Soc. Brasil. Mat. 13(1982), 3-17.

T. Januszkiewicz, Characteristic invariants of noncompact Riemannian manifolds, Topology 23(1984), 289-301.

S. Jekel, On two theorems of A. Haefliger concerning foliations, Topology 15(1976), 267-271.

S. Jekel, Simplicial K(G,1)'s, Manuscr. Math. 21(1977), 189-203.

S. Jekel, A note on the perfection of the fundamental group of the classifying space for codimension one real analytic foliations, Bol. Soc. Mat. Mex. 22(1977), 58-59.

S. Jekel, Loops on the classifying space for foliations, Amer. J. Math. 102(1980), 13-23.

S. Jekel, Simplicial decomposition of Γ-structures, Bol. Soc. Mat. Mex. 26(1981), 13-20.

D. Johnson, and A. Naveira, A topological obstruction to the geodesibility of a foliation of odd dimension, Geom. Dedicata 11(1981), 347-357.

D. Johnson and L. Whitt, Totally geodesic foliations on 3-manifolds, Proc. Amer. Math. Soc. 76(1979), 355-357.

D. Johnson and L. Whitt, Totally geodesic foliations, J. Diff. Geom. 15(1980), 225-235.

J. Jouanolou, Feuilles compactes des feuilletages algébriques, Math. Ann. 241(1979), 69-72.

G. Joubert, Estructuras foliadas, Rev. Colomb. Mat. 2(1968), 105-116.

G. Joubert and R. Moussu, El teorema de Novikov, Rev. Colomb. Mat. 2(1968), 117-123.

G. Joubert and R. Moussu, Feuilletage sans holonomie d'une variété fermée, C. R. Acad. Sci. Paris 270(1970), 507-509.

G. Joubert, R. Moussu, and D. Tischler, Sur les classes caractéristiques des feuilletages produits, C. R. Acad. Sci. Paris 275(1972), 171-174.

A. Justino, Some properties of the quotient manifold of a Jacobi manifold by a foliation generated by infinitesimal automorphisms, C. R. Acad. Sci. Paris 298(1984), 489-492.

A. Kabila, Sur les feuilletages logarithmiques, C. R. Acad. Sci. Paris 302(1986), 13-15.

F. Kamber, E. Ruh and Ph. Tondeur, Almost transversally symmetric foliations, Proc. of the II. Int. Symp. on differential geometry, Peniscola 1985, Springer Lecture Notes in 1209(1986), 184-189.

F. Kamber, E. Ruh and Ph. Tondeur, Comparing Riemannian foliations with transversally symmetric foliations, J. Diff. Geom. 27(1988).

F. Kamber and Ph. Tondeur, Invariant differential operators and the cohomology of Lie algebra sheaves, Memoirs Amer. Math. Soc. 113(1971), 1-125.

F. Kamber and Ph. Tondeur, Cohomologie des algèbres de Weil relatives tronquées, C. R. Acad. Sci. Paris 276(1973), 459-462.

F. Kamber and Ph. Tondeur, Algèbres de Weil semi-simpliciales, C. R. Acad. Sci. Paris 276(1973), 1177-1179.

F. Kamber and Ph. Tondeur, Homomorphisme caractéristique d'un fibre principal feuilleté, C. R. Acad. Sci. Paris 276(1973), 1407-1410.

F. Kamber and Ph. Tondeur, Classes caractéristiques dérivées d'un fibré principal feuilleté, C. R. Acad. Sci. Paris 276(1973), 1449-1452.

F. Kamber and Ph. Tondeur, Characteristic invariants of foliated bundles, Manuscr. Math. 11(1974), 51-89.

F. Kamber and Ph. Tondeur, Classes caractéristiques géneralisées des fibrés feuilletés localement homogenes, C. R. Acad. Sci. Paris 279(1974), 847-850.

F. Kamber and Ph. Tondeur, Quelques classes caractéristiques géneralisées non-triviales de fibrés feuilletés, C. R. Acad. Sci. Paris 279(1974), 921-924.

F. Kamber and Ph. Tondeur, Foliated bundles and characteristic classes, Springer Lecture Notes in Math. 494(1975).

F. Kamber and Ph. Tondeur, Semisimplicial Weil algebras and characteristic classes for foliated bundles in Cech cohomology, Proc. Symp. Pure Math., Stanford, Calif., 1973, Providence, vol. 27, Part 1 (1975), 283-294.

F. Kamber and Ph. Tondeur, Non-trivial characteristic invariants of homogeneous foliated bundles, Ann. Scient. Ec. Norm. Sup. 8(1975), 433-486.

F. Kamber and Ph. Tondeur, Classes caractéristiques et suites spectrales d'Eilenberg-Moore, C. R. Acad. Sci. Paris 283(1976), 883-886.

F. Kamber and Ph. Tondeur, G-foliations and their characteristic classes, Bull. Amer. Math. Soc. 84(1978), 1086-1124.

F. Kamber and Ph. Tondeur, On the linear independence of certain cohomology classes of BΓ, Advances in Math. Suppl. Studies 5(1979), 213-263.

F. Kamber and Ph. Tondeur, Feuilletages harmoniques, C. R. Acad. Sci. Paris 291(1980), 409-411.

F. Kamber and Ph. Tondeur, Harmonic foliations Proc. NSF Conference on Harmonic Maps, Tulane (1980), Springer Lecture Notes in Math. 949(1982), 87-121.

F. Kamber and Ph. Tondeur, Infinitesimal automorphisms and second variation of the energy for harmonic foliations, Tôhoku Math. J. 34(1982), 525-538.

F. Kamber and Ph. Tondeur, Dualité de Poincaré pour les feuilletages harmoniques, C. R. Acad. Sci. Paris 294(1982), 357-359.

F. Kamber and Ph. Tondeur, Duality for Riemannian foliations, Proc. Symp. Pure Math. vol. 40(1983); Part 1, 609-618.

F. Kamber and Ph. Tondeur, The index of harmonic foliations on spheres, Trans. Amer. Math. Soc. 275(1983), 257-263.

F. Kamber and Ph. Tondeur, Foliations and metrics, Proc. of a Year in Differential Geometry, University of Maryland, Birkhäuser, Progress in Mathematics vol. 32(1983), 103-152.

F. Kamber and Ph. Tondeur, Curvature properties of harmonic foliations, Illinois J. of Math. 18(1984), 458-471.

F. Kamber and Ph. Tondeur, Duality theorems for foliations, Astérisque 116(1984), 108-116.

F. Kamber and Ph. Tondeur, The Bernstein problem for foliations, Proc. of the Conference on Global Differential Geometry and Global Analysis, Berlin 1984, Springer Lecture Notes in Math. 1156(1985), 216-218.

F. Kamber and Ph. Tondeur, De Rham-Hodge theorem for Riemannian foliations, Math. Annalen 277(1987), 415-431.

F. Kamber, Ph. Tondeur and G. Toth, Transversal Jacobi fields for harmonic foliations, Mich. Math. J. 34(1987), 261-266.

Y. Kanie, Some Lie algebras of vector fields on foliated manifolds and their derivation algebras, Proc. Japan Acad. 55(1979), 409-411.

Y. Kanie, Cohomologies of Lie algebras of vector fields with coefficients in adjoint representations, Foliated case, Publs. Res. Inst. Math. Sci. 14(1978), 487-501.

W. Kaplan, Regular curve families filling the plane I, Duke Math. J. 7(1940), 154-185.

B. Kaup, Ein geometrisches Endlichkeitskriterium für Untergruppen von Aut(C, 0) und holomorphe 1-codimensionale Blätterungen, Comment. Math. Helv. 53(1978), 295-299.

H. Kitihara, The existence of complete bundle-like metrics, Ann. Sci. Kanazawa. Univ. 9(1972), 37-40.

H. Kitihara, The existence of complete bundle-like metrics, II, Ann. Sci. Kanazawa Univ. 10(1973), 51-54.

H. Kitihara, The completeness of a Clairaut foliation, Ann. Sci. Kanazawa Univ. 11(1974), 37-40.

H. Kitahara, Remarks on square-integrable basic cohomology spaces on a foliated Riemannian manifold, Kodai Math. J. 2(1979), 187-193.

H. Kitihara, Nonexistence of nontrivial harmonic 1-forms on a complete foliated Riemannian manifold, Trans. Amer. Math. Soc. 262(1980), 429-435.

H. Kitihara, On a parametrix form in a certain V-submersion, Springer Lecture Notes in Math. 792(1980), 264-298.

H. Kitihara, Stability theorems for Γ-foliations associated with semisimple flat homogeneous spaces, Ann. Sci. Kanazawa Univ. 21(1984), 7-18.

H. Kitihara and S. Yorozu, A formula for the normal part of the Laplace-Beltrami operator on the foliated manifold, Pacific J. Math. 69(1977), 425-432.

H. Kitihara and S. Yorozu, On some differential geometric characterization of a bundle-like metric, Kodai Math. J. 2(1979), 130-138.

U. Koschorke, Line fields transversal to foliations, Proc. Symp. Pure Math., Stanford, Calif., 1973, Providence, Vol. 27, Part 1 (1975), 295-301.

U. Koschorke, Bordism of plane fields, Proc. of the 9th Braz. Math. Colloq. 1973, Vol. II, Inst. Mat. Pura Apl. Sao Paolo 1977, 311-320.

Y. Kosmann-Schwarzbach, Lagrangian foliations and Lax equations, Lett. Math. Phys. 9(1985), 163-167.

J. Koszul, Homologie et cohomologie des algèbres de Lie, Bull. Soc. Math. France 78(1950), 65-128.

A. Kumpera and D. Spencer, Lie Equations, vol. I: General Theory, Ann. of Math. Studies, vol. 73, Princeton, NJ: Princeton Univ. (1972).

I. Kupka, The singularities of integrable structurally stable Pfaffian forms, Proc. Nat. Acad. Sci. USA 52(1964), 1431-1432.

J. Lacaze, Feuilletage d'une variété lorentzienne par des hypersurfaces spatiales à courbure moyenne constante, C. R. Acad. Sci. Paris 289(1979), 771-774.

C. Lamoureux, Feuilletages de codimension 1. Transversales fermées, C. R. Acad. Sci. Paris 270(1970), 1659-1662.

C. Lamoureux, Feuilletages de codimension 1. Holonomie et homotopie, C. R. Acad. Sci. Paris 270(1970), 1718-1721.

C. Lamoureux, Une condition pour qu'une feuille soit propre et ait une enveloppe composée de feuilles fermées, C. R. Acad. Sci. Paris 274(1972), 31-34.

C. Lamoureux, Feuilles fermées et captage; applications, C. R. Acad. Sci. Paris 277(1973), 579-581.

C. Lamoureux, Feuilles exceptionelles, feuilles denses, homologie et captage, C. R. Acad. Sci. Paris 277(1973), 1041-1043.

C. Lamoureux, Sur quelques phénomènes de captage, Ann. Inst. Fourier 23(1973), 229-243.

C. Lamoureux, The structure of foliations without holonomy on non-compact manifolds with fundamental group Z, Topology 13(1974), 219-224.

C. Lamoureux, Quelques propriétés globales des feuilletages par des feuilles simplement connexes non-nécessairement bornées, C. R. Acad. Sci. Paris 279(1974), 813-816.

C. Lamoureux, Groupes d'homologie et d'homologie d'orde supérieur des variétés compactes ou non compactes feuilletées en codimension 1, C. R. Acad. Sci. Paris 280(1975), 411-414.

C. Lamoureux, Non-bounded leaves in codimension one foliations, Springer Lecture Notes in Math. 484(1975), 257-272.

C. Lamoureux, Feuilles non captées et feuilles denses, Ann. Inst. Fourier 25(1975), 285-293.

C. Lamoureux, Feuilletages des variétés compactes et non compactes, Ann. Inst. Fourier 26(1976), 221-271.

C. Lamoureux, Geometric properties connected with the transverse structure of codimension one foliations, Astérisque 116(1984), 117-133.

R. Langevin, Courbure, feuilletages et singularités algébriques, Séminaire Le Dung Trang; Paris VII.

R. Langevin, Feuilletages tendus, Bulletin Soc. Math. France 107(1979), 271-281.

R. Langevin, Tight foliations, Exposé congrès Berlin Nov. (1979).

R. Langevin, Thèse d'Etat, Courbure, feuilletages et surfaces (mesures et distributions de Gauss) soutenue 1980 à l'Université de Paris XI - Orsay Publication mathématique d'Orsay N° 80-83.

R. Langevin, Feuilletages, énergies et cristaux liquides, Astérisque 107-108(1983), 201-213.

R. Langevin, Energie et géométrie intégrale, Peniscola (1982), Springer Lecture Notes in Math 1045(1984), 95-103.

R. Langevin and G. Levitt, Courbure totale des feuilletages des surfaces, Comment. Math. Helv. 57(1982), 175-195.

R. Langevin and G. Levitt, Sur la courbure totale des feuilletages des surfaces à bord, Bol. Soc. Brasil. Mat. 16(1985), 1-13.

R. Langevin and H. Rosenberg, On stability of compact leaves and fibrations, Topology 16(1977), 107-111.

R. Langevin and H. Rosenberg, Integrable perturbations of fibrations and a theorem of Seifert, Springer Lecture Notes in Math. 652(1978), 122-127.

F. Laudenbach and R. Roussarie, Un exemple de feuilletage sur S^3, Topology 9(1970), 63-70.

H. Lawson, Codimension-one foliations of spheres, Bull. Amer. Math. Soc. 77(1971), 437-438.

H. Lawson, Codimension-one foliations of spheres, Ann. Math. 94(1971), 494-503.

H. Lawson, Foliations, Bull. Amer. Math. Soc. 80(1974), 369-418.

H. Lawson, Lectures on the quantitative theory of foliation, CBMS Regional Conf. Series, Vol. 27(1977).

C. Lazarov, A permanence theorem for exotic classes, J. Diff. Geom. 14(1979), 475-486.

C. Lazarov, A vanishing theorem for certain rigid classes, Mich. Math. J. 28(1981), 89-95.

C. Lazarov, Characteristic classes for flat Diff(M)-bundles, preprint.

C. Lazarov, A note on the construction of elements in $H^*(\overline{B\,Diff(M)})$, preprint.

C. Lazarov, A relation between index and exotic classes, preprint.

C. Lazarov, An index theorem for foliations, Ill. J. of Math. 30(1986), 101-121.

C. Lazarov, Spectral invariants for foliations, Mich. Math. J. 33(1986), 231-243.

C. Lazarov and J. Pasternack, Secondary characteristic classes for Riemannian foliations, J. Diff. Geom. 11(1976), 365-385.

C. Lazarov and J. Pasternack, Residues and characteristic classes for Riemannian foliations, J. Diff. Geom. 11(1976), 599-612.

C. Lazarov and H. Shulman, Obstructions to foliation preserving Lie group actions, Topology 18(1979), 255-256.

C. Lazarov and H. Shulman, Obstructions to foliation-preserving vector fields, J. Pure Appl. Algebra 24(1982), 171-178.

D. Lehmann, J-homotopie dans les espaces de connexions et classes exotiques de Chern-Simons, C. R. Acad. Sci. Paris 275(1972), 835-838.

D. Lehmann, Classes caractéristiques exotiques et J-connexité des espaces de connexions, Ann. Inst. Fourier 24(1974), 267-300.

D. Lehmann, Sur l'approximation de certains feuilletages nilpotents par des fibrations, C. R. Acad. Sci. Paris 286(1978), 251-254.

D. Lehmann, Sur la généralization d'un théorème de Tischler à certains feuilletages nilpotents, Proc. K. Ned. Akad. Wet. Ser. A 82(1979), 177-189.

D. Lehmann, Structures de Maurer-Cartan et Γ_θ-structures, II, Espaces classifiants, Astérisque 116(1984), 134-138.

D. Lehmann, Modèle minimal relatif des feuilletages, Springer Lecture Notes in Math. 1183(1986), 250-258.

D. Lehmann and G. Martinez, Classes caracteristicas exoticas. Tipo de homotopia racional y formas differenciales, Gac. Mat. 26(1974), 147-157.

J. Lehmann-Lejeune, Cohomology over the transverse bundle to a foliation, C. R. Acad. Sci. Paris 295(1982), 495-498.

J. Lehmann-Lejeune, Cohomologies sur le fibré transverse à un feuilletage, Astérisque 116(1984), 149-179.

J. Leslie, A remark on the group of automorphisms of a foliation having a dense leaf, J. Diff. Geom. 7(1972), 597-601.

H. Levine and M. Shub, Stability of foliations, Trans. Amer. Math. Soc. 184(1973), 419-437.

G. Levitt, Feuilletages des variétés de dimension 3 qui sont des fibrés en cercles, Comment. Math. Helv. 53(1978), 572-594.

G. Levitt, Sur les mesures transverses invariantes d'un feuilletage de codimension 1, C. R. Acad. Sci. Paris 290(1980), 1139-1140.

M. Lewkowicz, On the action of diffeomorphisms on the Gelfand-Fuks cohomology, Bull. Polish Acad. Sci. Math. 32(1984), 695-701.

C. Liang, Multifoliations on open manifolds, Math. Ann. 221(1976), 143-146.

C. Liang, On the volume preserving foliations, Math. Ann. 223(1976), 13-17.

P. Libermann, Pfaffian systems and transverse differential geometry. Differential Geometry and Relativity: D. Reidel (1976), 107-126.

P. Libermann, Symplectic regular foliations, Atti Accad. Sci. Torino Cl. Sci. Fis. Mat. Natur. 117(1983), 239-246.

P. Libermann, Submanifolds and regular symplectic foliations, Pitman Research notes in Math. 80(1983), 81-106.

A. Lichnerowicz, Sur l'algèbre de Lie des champs de vecteurs, Comment. Math. Helv. 51(1976), 343-368.

A. Lichnerowicz, Feuilletage et algèbres de Lie associées, C. R. Acad. Sci. Paris 286(1978), 1141-1145.

A. Lichnerowicz, Algèbres de Lie attachées à un feuilletage, Ann. Fac. Sci. Toulouse Math 1(1979), 45-76.

A. Lichnerowicz, Feuilletage et déformations infinitésimales des algèbres de Lie associées, Proc. IV Int. Colloq. on Diff. Geom. Santiago de Compostela (1978), 192-204.

A. Lichnerowicz, Variétés de Poisson et feuilletages, Ann. Fac. Sci. Toulouse Math. 4(1982), 195-262.

A. Lichnerowicz, Formes caractéristiques d'un feuilletage et classes de cohomologie de l'algèbre des vecteurs tangents à valeurs dans les formes normales, C. R. Acad. Sci. Paris 296(1983), 67-71.

A. Lichnerowicz, Feuilletages, géométrie riemannienne et géométrie symplectique, C. R. Acad. Sci. Paris 296(1983), 205-210.

W. Lickorish, A foliation for 3-manifolds, Ann. of Math. 82(1975), 414-420.

L. Lininger, Codimension 1 foliations on manifolds with even index, Springer Lecture Notes in Math. 438(1975), 336-338.

T. Lokot, On foliations of codimension 1 with a trivial holonomy group, Usp. Mat. Nauk, 27(1972), 177-178.

T. Lokot, A topological classification of foliations of codimension 1 with trivial holonomy groups of class C^2 on the three-dimensional torus, Usp. Mat. Nauk, 27(1972), 205.

T. Lokot, On some properties of foliations of codimension 1 with trivial holonomy groups, in: Tr. Mosk. Inzh.-Stroit. Inst. 121(1974), 132-139.

B. Malgrange, Frobenius avec singularités I. Codimension un. Inst. Hautes Etudes Sci. Publ. Math. 46(1976), 163-173.

B. Malgrange, II. Le cas général, Invent. Math. 39(1977), 67-89.

R. Marinosci, Pseudoconnections on foliated manifolds, Note Mat. 1(1981), 71-92.

J. Marsden and F. Tipler, Maximal hypersurfaces and foliations of constant mean curvatue in general relativity, Phys. Reports 66(1980), 109-139.

J. Martinet, Classes caractéristiques des systèmes de Pfaff, Springer Lecture Notes in Math. 392(1974), 30-36.

X. Masa, Quelques propriétés des feuilletages de codimension 1 a connexion transverse projectable, C. R. Acad. Sci. Paris 284(1977), 811-812.

X. Masa, Cohomology of Lie foliations, Differential Geometry, Proceedings of the Vth Colloquium on Differential Geometry, Santiago de Compostela (1984), Pitman Research Notes 131(1986), 211-214.

J. Mather, The vanishing of the homology of certain groups of diffeomorphisms, Topology 10(1971), 297-298.

J. Mather, On Haefliger's classifying space, I, Bull. Amer. Math. Soc. 77(1971), 1111-1115.

J. Mather, On Haefliger's classifying space, II, Harvard University.

J. Mather, Integrability in codimension 1, Comment. Math. Helv. 48(1973), 195-233.

J. Mather, Foliations and homology of groups of diffeomorphisms, Proc. Int. Congr. Math., Vancouver 1974, vol. 2,S. 1(1975), 35-37.

J. Mather, Foliations of surfaces, I, An ideal boundary, Ann. Inst. Fourier 32(1982), 235-261.

T. Matsuoka and S. Morita, On characteristic classes of Kähler foliations, Osaka J. Math. 16(1979), 539-550.

J. Mattei, Holonomie et intégrales premières, Ann. Scien. Ec. Norm. Sup. 4° S.,t. 12(1980), 469-523.

D. McDuff, Foliations and monoids of embeddings, Proc. Georgia Conf. on Geometric Topology 1977, 429-444.

D. McDuff, On groups of volume preserving diffeomorphisms, and foliations with transverse volume form, Proc. London Math. Soc. 43(1981), 295-320.

A. Medina, Nombres caractéristiques du fibré transverse à un feuilletage et champs de vecteurs feuilletés, C. R. Acad. Sci. Paris 276(1973), 863-865.

A. Medina, Quelques remarques sur les feuilletages affines, C. R. Acad. Sci. Paris 282(1976), 1159-1162.

A. Medina, Quelques remarques sur les feuilletages affines, Cah. Topol. Geom. Different., Ch. Ehresmann 17(1976), 59-68.

X. Mei, Note on the residues of the singularities of a Riemannian foliation, Proc. Amer. Math. Soc. 89(1983), 359-366.

J. Meyer, e-foliations of codimension two, J. Diff. Geom. 12(1977), 583-594.

I. Mihut, Chern classes of foliate manifolds, Inst. Politehn. Traian Vuia Timisoara. Lucrar. Sem. Mat. Fiz. (1983), 29-32.

K. Millet, Compact foliations, Springer Lecture Notes in Math. 484(1975), 277-287.

K. Millet, Generic properties of proper foliations, I.H.E.S. preprint.

M. Miniconi, Familles de feuilletages analytiques, C. R. Acad. Sci. Paris 278(1974), 273-276.

N. Mishachev, Elimination of the standard singularities of flags of foliations, Syktyvkar Univ. (1978).

N. Mishachev, The construction of flags of foliations, Usp. Mat. Nauk. 34(1979), 237-238.

N. Mishachev and Y. Éliashberg, Survey on singularities of foliations, Funkts. Anal. Ego Prilozhen 11(1977), 43-53.

S. Miyoshi, On the placement problem of Reeb components, Comment. Math. Helv. 57(1982), 260-281.

S. Miyoshi, Foliated round surgery of codimension-one foliated manifolds, Topology 21(1982), 245-261.

T. Mizutani, Remarks on codimension one foliations of spheres, J. Math. Soc. Japan. 24(1972), 732-735.

T. Mizutani, Foliated cobordisms of S^3 and examples of foliated 4-manifolds, Topology 13(1974), 353-362.

T. Mizutani, Foliations and foliated cobordisms of spheres in codimension one, J. Math. Soc. Japan. 27(1975), 264-280.

T. Mizutani, S. Morita, and T. Tsuboi, The Godbillon-Vey class of codimension one foliations which are almost without holonomy, Ann. of Math. 113(1981), 515-527.

T. Mizutani, S. Morita and T. Tsuboi, On the cobordism classes of codimension one foliations which are almost without holonomy, Topology 22(1983), 325-343.

T. Mizutani and T. Tsuboi, Foliations without holonomy and foliated bundles, Sci. Rep. Saitama Univ. 9(1979), 45-55.

N. Mok, Foliation techniques and vanishing theorems, Contemporary Mathematics 49(1986), 79-118.

P. Molino, Connexions et G-structures sur les variétés feuilletées, Bull. Sci. Math. 92(1968), 59-63.

P. Molino, G-structures plates et classes caractéristiques, C. R. Acad. Sci. Paris 269(1969), 917-919.

P. Molino, Classe d'Atiyah d'un feuilletage et connexions transverses projetables, C. R. Acad. Sci. Paris 272(1971), 779-781.

P. Molino, Classes caractéristiques et obstructions d'Atiyah pour les fibrés principaux feuilletés, C. R. Acad. Sci. Paris 272(1971), 1376-1378.

P. Molino, Feuilletages et classes caractéristiques, Symp. Math. Conv., 1971-1972, London-New York, Vol, 10(1972), 119-209.

P. Molino, Propriétés cohomologiques et propriétés topologiques des feuilletages à connexions transverses projetables, Topology 12(1973), 317-325.

P. Molino, La classe d'Atiyah d'un feuilletage comme cocycle de déformation infinitésimale, C. R. Acad. Sci. Paris 278(1974), 719-721.

P. Molino, Γ_q-structures partielles et classes de Bott-Haefliger, C. R. Acad. Sci. Paris 281(1975), 203-206.

P. Molino, Sur la géométrie transverse des feuilletages, Ann. Inst. Fourier 25(1975), 279-284.

P. Molino, Feuilletages transversement parallélisables et feuilletages de Lie, Applications, C. R. Acad. Sci. Paris 282(1976), 99-101.

P. Molino, Etude des feuilletages transversalement complets et applications, Ann. Sci. Ecole Norm. Super. 10(1977), 289-307.

P. Molino, Feuilletages Riemanniens sur les variétés compactes; champs de Killing transverses, C. R. Acad. Sci. Paris 289(1979), 421-423.

P. Molino, Géometrie globale des feuilletages riemanniens, Proc. Kon. Nederland Akad. Ser. A, 1, 85(1982), 45-76.

P. Molino, Feuilletages riemanniens, Secrétariat des Mathématiques, Université des Sciences et Technique du Languedoc (1982-1983).

P. Molino, Flots riemanniens et flots isométriques, Séminaire de géometrie différentielle (1982-83), Montpellier.

P. Molino, Espace des feuilles des feuilletages riemanniens, Astérisque 116(1984), 180-189.

P. Molino, Désingularisation des feuilletages riemanniens, Amer. J. Math. 106(1984), 1091-1106.

P. Molino, Feuilletages riemanniens réguliers et singuliers, to appear.

P. Molino et V. Sergiescu, Deux remarques sur les flots riemanniens, Manuscripta Math. 51(1985), 145-161.

A. Montesinos, On certain classes of almost product structures, Mich. Math. J. 30(1983).

A. Montesinos, Conformal curvature for the normal bundle of a conformal foliation, Ann. Inst. Fourier 32(1982), 261-274.

A. Morgan, Holonomy and metric properties of foliations in higher codimension, Proc. Amer. Math. Soc. 58(1976), 255-261.

A. Morimoto, Prolongations of geometric structures, Lecture Notes Math. Inst. Nagoya Univ. 1969.

S. Morita, A remark on the continuous variations of secondary characteristic classes for foliations, J. Math. Soc. Japan. 29(1977), 253-260.

S. Morita, Cartan connections and characteristic classes of foliations, Proc. Japan. Acad. 53(1977), 211-214.

S. Morita, On characteristic classes of Riemannian foliations, Osaka J. Math. 16(1979), 161-172.

S. Morita, On characteristic classes of conformal and projective foliations, J. Math. Soc. Japan 31(1979), 693-718.

S. Morita, On the splitting problem of the boundary functionals for the characteristic classes of foliations, Topology 20(1981), 411-420.

S. Morita, Nontriviality of the Gelfand-Fuks characteristic classes for flat S^2-bundles, Osaka J. Math. 21(1984), 545-563.

S. Morita and T. Tsuboi, The Godbillon-Vev class of codimension one foliations without holonomy, Topology 19(1980), 43-49.

I. Moskowitz, Nonvanishing local cohomology classes, Trans. Amer. Math. Soc. 286(1984), 831-837.

I. Moskowitz, A note on the Bott vanishing theorem, Proc. Amer. Math. Soc. 94(1985), 529-530.

J. Morvan, Connection and foliation in symplectic geometry, C. R. Acad. Sci. Paris 296(1983), 765-768.

M. Mostow, Continuous cohomology of spaces with two topologies, Mem. Amer. Math. Soc. 175(1976).

M. Mostow, Variations, characteristic classes, and the obstruction to mapping smooth to continuous cohomology, Trans. Amer. Math. Soc. 240(1978), 163-182.

R. Moussu, Sur un théorème de Novikov, Rev. Colomb. Mat. 3(1969), 51-81.

R. Moussu, Feuilletage sans holonomie d'une variété fermée, C. R. Acad. Sci. Paris 270(1970), 1308-1311.

R. Moussu, Feuilletage de codimension 1 transverse au bord, C. R. Acad. Sci. Paris 271(1970), 15-18.

R. Moussu, Feuilletages presque sans holonomie, C. R. Acad. Sci. Paris 272(1971), 114-117.

R. Moussu, Sur les problèmes de classification de feuilletages, Rev. Colomb. Mat. 6(1972), 59-68.

R. Moussu, Sur les classes exotiques des feuilletages, Springer Lecture Notes in Math. 392(1974), 37-42.

R. Moussu, Holonomie évanescente des équations différentielles dégénerées transverses, Singularities and dynamical systems, Iráklion, 1983; North-Holland Math. Studies 103(1985), 161-173.

R. Moussu and F. Pelletier, Sur le théorème de Poincaré-Bendixson, Ann. Inst. Fourier 24(1974), 131-148.

R. Moussu and R. Roussarie, Une condition suffisante pour qu'un feuilletage soit sans holonomie, C. R. Acad. Sci. Paris 271(1970), 240-243.

M. Muller, Quelques propriétés des feuilletages polynomiaux du plan, Bol. Soc. Mat. Mex. 21(1976), 6-14.

M. Muller, An analytic foliation of the plane without weak first integrals of class C^2, Bol. Soc. Mex. 21(1976), 1-5.

M. Muller, Sur les composantes de Novikov des feuilletages, Topology 19(1980), 199-201.

T. Müller, Beispiel einer periodischen instabilen holomorphen Strömung, Enseign. Math. 25(1979), 309-312.

H. Münzner, Isoparametrische Hyperflächen in Sphären, Math. Ann. 251(1980), 57-71.

H. Münzner, Isoparametrische Hyperflächen in Sphären, II, Über die Zerlegung der Sphäre in Ballbündel, Math. Ann. 256(1981), 215-232.

A. Narmanov, Limit sets of leaves of a foliation of codimension 1, Vestnik Leningrad. Univ. Mat. Mekh. Astronom. 3(1983), 21-25.

A. Narmanov, Controllability sets of control systems that are fibers of a foliation of codimension one, Differentsial'nye Uravneniya 19(1983), 1627-1630.

S. Nagamine, A remark on minimal foliations of Lie groups, Tsukuba J. Math. 9(1985), 317-320.

H. Natsume and T. Natsume, A remark on the DeRham map for foliated manifolds, Kodai Math. J. 3(1980), 364-373.

H. Natsume and T. Natsume, On a theorem of Bott and Haefliger, Sci. Rep. Saitama Univ. Ser. A 9(1980), 81-95.

A. Naveira, Variedades foliadas con metrica casi-fibrada, Collect. Math. 21(1970), 5-61.

A. Naveira, A classification of Riemannian almost product manifolds, Rend. Math. 3(1983), 577-592.

A. Naveira and A. Rocamora, A geometric obstruction to the existence of two totally umbilic complementary foliations in compact manifolds, Springer Lecture Notes in Mathematics 1139(1985), 263-279.

M. Nicolau and A. Reventos, Compact Hausdorff foliations, I, Int. Symp. on differential geometry, Peniscola 1982, Springer Lecture Notes in Math. 1045(1984), 147-153.

S. Nishikawa, Residues and secondary characteristic classes for projective foliations, Proc. Japan. Acad. 54(1978), 79-82.

S. Nishikawa, Residues and characteristic classes for projective foliations, Japan J. Math. 7(1981), 45-108.

S. Nishikawa and H. Sato, On characteristic classes of Riemannian, conformal and projective foliations, J. Math. Soc. Japan. 28(1976), 224-241.

S. Nishikawa and M. Takeuchi, Γ-foliations and semisimple flat homogeneous spaces, Tôhoku Math. J. 30(1978), 307-335.

S. Nishikawa and Ph. Tondeur, Transversal infinitesimal automorphisms for harmonic Kähler foliations, to appear.

T. Nishimori, Isolated ends of open leaves of codimension-one foliations, Q. J. Math. 26(1975), 159-167.

T. Nishimori, Compact leaves with Abelian holonomy, Tôhoku Math. J. 27(1975), 259-272.

T. Nishimori, Behaviour of leaves of codimension one foliations, Tôhoku Math. J. 29(1977), 255-273.

T. Nishimori, Ends of leaves of codimension-one foliations, Tôhoku Math. J. 31(1979), 1-22.

T. Nishimori, SRH-decompositions of codimension-one foliations and the Godbillon-Vey classes, Tôhoku Math. J. 32(1980), 9-34.

T. Nishimori, Existence problem of transverse foliations for some foliated 3-manifolds, Tôhoku Math. J. 34(1982), 179-238.

T. Nishimori, Foliations transverse to the turbulized foliations of punctured torus bundles over a circle, Hokkaido Math. J. 13(1984), 1-25.

J. Noakes, Foliations by manifolds with boundaries, J. Diff. Geom. 16(1981), 129-136.

S. Novikov, Topology of foliations, Trudy Moskov. Mat. Obsc. 14(1965), 248-278; AMS Translation, Trans. Moscow Math. Soc. 14(1967), 268-304.

H. Ohsato, Characteristic classes of S-foliated bundles and Gelfand-Fuks cohomology, J. Fac. Sci. Univ. Tokyo 26(1979), 279-301.

B. O'Neill, The fundamental equations of a submersion, Mich. Math. J. 13(1966), 459-469.

G. Oshikiri, The surgery of codimension-one foliations, Tôhoku Math. J. 31(1979), 63-70.

G. Oshikiri, A remark on minimal foliations, Tôhoku Math. J. 33(1981), 133-137.

G. Oshikiri, Jacobi fields and the stability of leaves of codimension-one minimal foliations, Tôhoku Math. J. 34(1982), 417-424.

G. Oshikiri, Totally geodesic foliations and Killing fields, Tôhoku Math. J. 35(1983), 387-392.

G. Oshikiri, Totally geodesic foliations and Killing fields, II, Tôhoko Math. J. 38(1986), 351-356.

G. Osipenko, Integrability of invariant plane fields, separation and partial hyperbolicity, Differential Equations 19(1983), 1251-1255.

M. O'uchi, Coverings of foliations and associated C^*-algebras, Math. Scand. 58(1986), 69-76.

R. Palais, A global formulation of the Lie Theory of Transformation Groups, Mem. Amer. Math. Soc. 22(1957).

J. Palis, Rigidity of the centralizers of diffeomorphisms and structural stability of suspended foliations, Springer Lecture Notes in Math. 652(1978), 114-121.

C. Palmeira, Variétés ouvertes feuilletées par plans, C. R. Acad. Sci. Paris 283(1976), 237-239.

C. Palmeira, Open manifolds foliated by planes, Ann. of Math. 107(1978), 109-131.

C. Palmeira, Feuilletages par cylindres fermés de $\mathbb{R}^3$, C. R. Acad. Sci. Paris 290(1980), 419-421.

C. Palmeira, Erratum Feuilletages par cylindres femés de $\mathbb{R}^3$, C. R. Acad. Sci. Paris 290(1980), 929.

C. Palmeira, Foliations by closed cylinders in 3-dimensional manifolds, Bol. Soc. Brasil. Mat. 13(1982), 55-78.

Ph. Parker, Geometry of leaves and the heat equation, Teubner Texte zur Mathematik 57(1983), 247-251.

J. Pasternack, Topological obstructions to integrability and Riemannian geometry of foliations, Thesis, Princeton University, Princeton (1970).

J. Pasternack, Foliations and compact Lie groups actions, Comment. Math. Helv. 46(1971), 467-477.

J. Pasternack, Classifying spaces for Riemannian foliations, Proc. Symp. Pure Math., Stanford, Calif. (1973), Providence, vol. 27, Part 1 (1975), 303-310.

W. Pelletier, The secondary characteristic classes of solvable foliations, Proc. Amer. Math. Soc. 88(1983), 651-659.

J. Perchik, Cohomology of Hamiltonian and related formal vector field Lie algebras, Topology 15(1976), 395-404.

A. Phillips, Submersions of open manifolds, Topology 6(1967), 171-206.

A. Phillips, Foliations of open manifolds, I, Comment. Math. Helv. 43(1968), 204-211.

A. Phillips, Foliations of open manifolds, II, Comment. Math. Helv. 44(1969), 367-370.

A. Phillips, Smooth maps transverse to a foliation, Bull. Amer. Math. Soc. 76(1970), 792-797.

A. Phillips and D. Stone, The Euler cycle of a foliation, J. Diff. Geom. 15(1980), 39-50.

A. Phillips and D. Sullivan, Geometry of leaves, Topology 20(1981), 209-218.

P. Piccinni, A formula for the trace Laplacian of the second fundamental form of a foliation, to appear.

M. Pierrot, Orbites des champs feuilletés pour un feuilletage riemannien sur une variété compacte, C.R.Acad. Sci. Paris 301(1985), 443-445.

G. Pitis, On the cohomology of foliate manifolds, Bul. Univ. Brasov. Ser. C 22(1980), 101-106.

G. Pitis, A class of chain complexes of a foliated manifold, An. Stiint. Univ. Al. I. Cuza Iaci Sect. I a Mat. 27(1981), 63-66.

G. Pitis, A cohomology theory on the category of foliated manifolds, Studia Univ. Babes-Bolyai Math. 29(1984), 33-38.

H. Pittie, The secondary characteristic classes of parabolic foliations, Comment. Math. Helv. 54(1979), 601-614.

J. Plante, Asymptotic properties of foliations, Comment. Math. Helv. 47(1972), 449-456.

J. Plante, A generalization of the Poincaré-Bendixson theorem for foliation of codimension one, Topology 12(1973), 177-181.

J. Plante, On the existence of exceptional minimal sets in foliations of codimension one, J. Diff. Eq. 15(1974), 178-194.

J. Plante, Foliations transverse to fibers of a bundle, Proc. Amer. Math. Soc. 24(1974), 631-635.

J. Plante, Foliations with measure preserving holonomy, Springer Lecture Notes in Math. 468(1975), 6-7.

J. Plante, Foliations with measure preserving holonomy, Ann. of Math. 102(1975), 327-361.

J. Plante, Foliations of 3-manifolds with solvable fundamental group, Invent. Math. 51(1979), 219-230.

J. Plante, Anosov flows, transversely affine foliations and a conjecture of Verjovsky, J. London Math. Soc. 23(1981), 359-362.

J. Plante, Stability of codimension one foliations by compact leaves, Topology 22(1983), 173-177.

J. Plante and W. Thurston, Polynomial growth in holonomy groups of foliations, Comment. Math. Helv. 51(1976), 567-584.

V. Podolski and A. Sirokov, Some types of Riemannian foliations, Gravitacija i Teor. Otnositelnosti 10-11(1975), 232-236.

J. Pradines, Remarque sur le théorème d'annulation de Bott-Martinet, C. R. Acad. Sci. Paris 282(1976), 527-529.

J. Pradines, Feuilletages dout les groupes d'holonomie sont de Coxeter, C. R. Acad. Sci. Paris 286(1978), 255-258.

J. Pradines, Un feuilletage sans holonomie transversale, dont le quotient n'est pas une Q-variété, C. R. Acad. Sci. Paris 288(1979), 245-248.

J. Pradines, Foliations: holonomy and local graphs, C. R. Acad. Sci. Paris 298(1984), 297-300.

J. Pradines, How to define the differentiable graph of a singular foliation, Cahiers Topol. Géom. Différentielle Catégoriques 26(1985), 339-380.

J. Pradines and B. Bigonnet, Graphe d'un feuilletage singulier, C. R. Acad. Sci. Paris 300(1985), 439-442.

J. Pradines and J. Wuafo Kanga, Relations d'équivalence transversalement différentiables, C. R. Acad. Sci. Paris 283(1976), 25-28.

M. Puta, Some remarks on the relative cohomology of a foliated manifold, An. Univ. Timişoara Ser. Stiint. Mat. 23(1985), 56-60.

Ngo van Que, Feuilletage à singularités de variétés de dimension 3 (théorème de J. Wood), J. Diff. Geom. 6(1972), 473-478.

Ngo van Que, Formes singulières génériques, Proc. 9th Braz. Math. Coll. 1973, Vol. I, Inst. Mat. Pura Apl. Sao Paulo (1977), 217-226.

A. Ranjan, Structural equations and integral formula for foliated manifolds, Geometriae Dedicata 20(1986), 85-91.

O. Rasmussen, Foliations with bundle-like metric, Prepr. Ser. Mat. Inst. Aarhus Univ. No. 26, 33(1971-1972).

O. Rasmussen, Locally free R^{n-1} actions on M^n, Prepr. Ser. Mat. Inst. Aarhus Univ. No. 32, 13(1971-1972).

O. Rasmussen, Reeb foliations, Prepr. Ser. Mat. Inst. Aarhus Univ. No. 58, 8(1973).

O. Rasmussen, The horocyclic foliation, Prepr. Ser. Mat. Inst. Aarhus Univ. No. 3, 21(1975-1976).

O. Rasmussen, Exotic characteristic classes for holomorphic foliations, Invent. Math. 46(1978), 153-171.

O. Rassmussen, Continuous variation of foliations in codimension two, Topology 19(1980), 335-349.

O. Rasmussen, Foliations with integrable transverse G-structures, J. Diff. Geom. 16(1981), 699-710.

B. Raymond, Ensembles de Cantor et feuilletages, Thèse Doct. Sci. Univ. Paris, (1976).

C. Rea, Levi-flat submanifolds and holomorphic extension of foliations, Ann. Suola Norm. Super. Pisa., Sci. Fis. Mat. 26(1972) 665-681.

C. Rea, Varieta pseudo-platte, Symp. Math. 1st. Naz. Alta Mat. Conv., Nov. 1971-Maggio 1972, London-New York, vol. 11(1973), 347-354.

G. Reeb, Variétés feuilletées, feuilles voisines, C. R. Acad. Sci. Paris 224(1947), 1613-1614.

G. Reeb, Sur certaines propriétés topologiques des variétés feuilletées, Actualité Sci. Indust. 1183, Hermann, Paris (1952).

G. Reeb, Sur les structures feuilletées de codimension un et sur un théorème de M. A. Denjoy, Ann. Inst. Fourier 11(1961), 185-200.

G. Reeb, Structures feuilletées, Springer Lecture Notes in Math. 652(1978), 104-113.

B. Reinhart, Harmonic integrals on almost product manifolds, Trans. Amer. Math. Soc. 88(1958), 243-276.

B. Reinhart, Foliated manifolds with bundle-like metrics, Ann. of Math. 69(1959), 119-132.

B. Reinhart, Harmonic integrals on foliated manifolds, Amer. J. Math. 81(1959), 529-536.

B. Reinhart, Line elements on the torus, Amer. J. Math. 81(1959), 617-631.

B. Reinhart, Closed metric foliations, Mich. Math. J. 8(1961), 7-9.

B. Reinhart, Structures transverse to a vector field, Int. Symp. on nonlinear differential equations and nonlinear mechanics, Academic Press, New York (1963), 442-444.

B. Reinhart, Cobordism and foliations, Ann. Inst. Fourier 14(1964), 49-52.

B. Reinhart, Characteristic numbers of foliated manifolds, Topology 6(1967), 467-472.

B. Reinhart, Algebraic invariants of foliations, Symposium on Differential Equations and Dynamical Systems 1968-69, University of Warwick, Berlin, Heidelberg, Springer-Verlag, New York (1971), 119-120.

B. Reinhart, Automorphisms and integrability of plane fields, J. Diff. Geom. 6(1971), 263-266.

B. Reinhart, Indices for foliations of the two dimensional torus, Symposium Dynamical Systems, Salvador, 1971, 421-424; New York, Academic Press, (1973).

B. Reinhart, Maximal foliations of extended Schwarzschild space, J. Math. Phys. 14(1973), 719.

B. Reinhart, Holonomy invariants for framed foliations, Colloque de Géométrie Différentielle, Santiago de Compostela, 1972, 47-52; Berlin, Heidelberg, New York, Springer-Verlag, (1974).

B. Reinhart, Foliation invariants and leaves, Holiday Symposium on Foliations and the Gelfand-Fuks cohomology, Las Cruces, 11 pages; New Mexico State University (1975).

B. Reinhart, The second fundamental form of a plane field, J. Diff. Geom. 12(1977), 619-627.

B. Reinhart, Foliations and second fundamental form, Fourth colloquium on differential geometry, Santiago de Compostela, Spain (1978), 246-253.

B. Reinhart, Differential geometry of foliations, Ergeb. Math. 99(1983), Springer-Verlag, New York.

B. Reinhart and J. Wood, A metric formula for the Godbillon-Vey invariant for foliations, Proc. Amer. Math. Soc. 38(1973), 427-430.

C. Remsing, Introducere in Teoria Geometrica a Foliatiilor (Romania) (1985).

A. Reznikov, Complete geodesic foliations of Lie groups (Russian), Differentsia'naya Geom. Mnogoobraz. Figur. 16(1985), 67-70, 124.

A. Rocamora, Some geometric consequences of the Weitzenböck formula on Riemannian almost-product manifolds. Harmonic distributions, to appear.

C. Roger, Sur les classes caractéristiques des feuilletages donnés par des isométries, C. R. Acad. Sci. Paris 276(1973), 1185-1188.

C. Roger, Etude des Γ-structures de codimension 1 sur la sphere S^2, Ann. Inst. Fourier 23(1973), 213-227.

C. Roger, Méthodes homotopiques et cohomologiques en théorie des feuilletages, Thèse Doct. Sc. Math., Univ. Paris (1976).

C. Roger, Homologies of affine groups and classifying spaces of polylinear foliations, Funkts. Anal. Ego Prilozhen 13(1979), 47-52, 96.

C. Roger, Sur la cohomologie de l'espace classifiant des feuilletages symplectiques, C. R. Acad. Sci. Paris, 290(1980), 617-619.

C. Roger, Foliations with a symplectic or contact transverse structure, Symplectic Geometry, Toulouse (1981), Pitman Research Notes in Math. 80(1983), 243-250.

C. Roger, Cohomologie (p,q) des feuilletages et applications, Astérisque 116(1984), 195-213.

H. Rosenberg, Feuilletages sur des sphères (d'après H. Blaine Lawson), Springer Lecture Notes in Math. 383(1974), 294-306.

H. Rosenberg, Labyrinths in the disc and surfaces, Ann. of Math. 117(1983), 1-33.

H. Rosenberg and R. Roussarie, Reeb foliations, Ann. Math. 91(1970), 1-24.

H. Rosenberg and R. Roussarie, Topological equivalence of Reeb foliations, Topology 9(1970), 231-242.

H. Rosenberg and R. Roussarie, Les feuilles exceptionelles ne sont pas exceptionelles, Comment. Math. Helv. 45(1970), 517-523.

H. Rosenberg and R. Roussarie, Some remarks on stability of foliations, J. Diff. Geom. 10(1975), 207-219.

R. Roussarie, Sur les feuilletages des variétés de dimension trois, Thèse Doct. Sci. Math., Fac. Sci. Orsay Univ., Paris, 70(1969).

R. Roussarie, Sur les feuilletages des variétés de dimension trois, Ann. Inst. Fourier 21(1971), 18-32.

R. Roussarie, Plongement dans les variétés feuilletées et classification de feuilletages sans holonomie, Inst. Hautes Etudes Sc. Publ. Math. 43(1973), 101-104.

R. Roussarie, Phénomènes de stabilité et d'instabilité dans les feuilletages, Springer Lecture Notes in Math. 392(1974), 53-60.

R. Roussarie, Constructions de feuilletages (d'après W. Thurston), Springer Lecture Notes in Math. 677(1978), 138-154.

V. Rovenskii, Totally geodesic foliations, Sibirsk. Mat. Zh. 23(1982), 217-219, 224.

V. Rubanov, Transversal and projectable invariant connections, Vestsi Akad. Navuk BSSR Ser. Fiz. Mat. Navuk (1983), 47-51.

V. Rubanov, Invariant foliations and transversal connections, Dokl. Akad. Nauk BSSR 28(1984), 696-697.

D. Ruelle, Integral representation of measures associated with a foliation, Inst. Hautes Etudes Sci. Publ. Math. 48(1978), 127-132.

D. Ruelle, Invariant measures for a diffeomorphism which expands the leaves of a foliation, Inst. Hautes Etudes Sci. Publ. Math. 48(1978), 133-135.

E. Ruh and J. Vilms, The tension field of the Gauss map, Trans. Amer. Math. Soc. 149(1970), 569-573.

H. Rummler, Quelques notions simples en géométrie riemannienne et leur applications aux feuilletages compacts, Comment. Math. Helv. 54(1979), 224-239.

H. Rummler, Kompakte Blätterungen durch Minimalflächen, Habilitationsschrift Universität Freiburg i.Ve. (1979).

M. Samuélidès, Tout feuilletage à croissance polynomiale est hyperfini, J. Funct. Anal. 34(1979), 363-369.

R. Sacksteder, On the existence of exceptional leaves in foliations of codimension one, Ann. Inst. Fourier 14(1964), 221-226.

R. Sacksteder, Foliations and pseudo-groups, Amer. J. Math. 87(1965), 79-102.

R. Sacksteder, A remark on Thurston's stability theorem, Ann. Inst. Fourier 25(1975), 219-220.

R. Sacksteder, Foliations and separation of variables, Astérisque 116(1984), 214-222.

R. Sacksteder, and A. Schwartz, Limit sets of foliations, Ann. Inst. Fourier 15(1965), 201-214.

E. Salhi, On local minimal sets, C. R. Acad. Sci. Paris 295(1982), 691-694.

E. Salhi, Sur un théorème de structure des feuilletages de codimension 1, C. R. Acad. Sci. Paris 300(1985), 635-638.

E. Salhi, Niveau des feuilles, C. R. Acad. Sci. Paris 301(1985), 219-222.

E. Sallum, Vector fields tangent to a Reeb foliation on S^3, J. Diff, Eq. 34(1979), 204-211.

J. Sampson, Foliations from quadratic and hermitian differential forms, Arch. Rat. Mech. and Anal. 70(1979), 91-99.

M. Saralegui, The Euler class for flows of isometries, Pitman Research Notes in Math. 131(1985), 220-227.

G. Sardanashvily, Space-time foliations, Acta Physica Hungarica 57(1985), 31-40.

G. Sardanashvily and V. Yauchevskii, Space-time foliations in the theory of gravitation, Izv. Vyss. Uchebn. Zaved. Fiz. (1982), 20-23.

K. Sarkaria, A finiteness theorem for foliated manifolds, J. Math. Soc. Japan. 30(1978), 687-696.

A. Sato, Stably solitary foliations, Hokkaido Math. J. 14(1985), 143-147.

S. Schecter and M. Singer, Planar polynomial foliations, Proc. Amer. Math. Soc. 79(1980), 649-656.

S. Schecter and M. Singer, Addendum to Planar polynomial foliations, Proc. Amer. Math. Soc. 83(1981), 220.

A. Schwartz, A generalization of the Poincaré-Bendixson theorem to closed two-dimensional manifolds, Amer. J. Math. 85(1963), 453-458.

S. Schwartzman, Asymptotic cycles, Ann. of Math. 66(1957), 270-284.

G. Schwarz, On the De Rham cohomology of the leaf space of a foliation, Topology 13(1974), 185-187.

P. Schweitzer, Counterexample to the Seifert conjecture and opening leaves of foliations, Ann. of Math. 100(1974), 386-400.

P. Schweitzer, Compact leaves of foliations, Springer Lecture Notes in Math. 468(1975), 4-6.

P. Schweitzer, Compact leaves of codimension one foliations, Springer Lecture Notes in Math. 484(1975), 273-276.

P. Schweitzer, Codimension one plane fields and foliations, Proc. Symp. Pure Math., Stanford, Calif., 1973, Providence, vol. 27, Part 1 (1975), 311-312.

P. Schweitzer, Compact leaves of foliations, Proc. Int. Congr. Math., Vancouver, 1974, vol. 1, S. 1(1975), 543-546.

P. Schweitzer (Ed), Some problems in foliation theory and related areas, Springer Lecture Notes in Math. 652(1978), 240-252.

P. Schweitzer, Stability of compact leaves with trivial linear holonomy, preprint.

P. Schweitzer and P. Whitman, Pontryagin polynomial residues of isolated foliation singularities, Springer Lecture Notes in Math. 652(1978), 95-103.

A. Sec, Sur certaines équations de Pfaff complètement intégrables dans le champ complexe (propriétés du feuilletage associé), Springer Lecture Notes in Math. 484(1975), 224-233.

A. Sec and R. Gerard, Feuilletages de Painlevé et équations de Pfaff, C. R. Acad. Sci. Paris 270(1970), 1166-1169.

G. Segal, Classifying spaces related to foliations, Topology 17(1978), 367-382.

A. Seitoh, Remarks on stability for semiproper exceptional leaves, Tokyo J. Math. 6(1983), 95-108.

B. Seke, Sur les structures transversalement affines des feuilletages de codimension un, Ann. Inst. Fourier 30(1920), 1-29.

B. Seka, Structures transverses affines trivialisables, Publ. IRMA Strasbourg 188 P-108(1982).

F. Sergeraert, Feuilletages et difféomorphismes infinitement tangent à l'identité, Invent. Math. 39(1977), 253-275.

F. Sergeraert, BΓ (d'après Mather et Thurston), Springer Lecture Notes in Math. 710(1979), 300-315.

V. Sergiescu, Cohomologie basique et dualité des feuilletages Riemanniens, Ann. Inst. Fourier 35(1985), 137-158.

K. Shibata, On Haefliger's model for the Gelfand-Fuks cohomology, Japan J. Math. 7(1981), 379-415.

Y. Shikata, On the cohomology of bigraded forms associated with foliated structures, Bull. Soc. Math. Grèce 15(1974), 68-76.

Y. Shikata, On a homology theory associated to foliations, Nagoya Math. J. 38(1970), 53-61.

H. Shulman, Characteristic classes and foliations, Thesis, University of California, Berkeley (1972).

H. Shulman, Secondary obstruction to foliations, Topology 13(1974), 177-183.

H. Shulman, The double complex of Γ_k, Proc. Symp. Pure Math., Stanford, Calif., 1973, Providence, vol. 27, Part 1 (1975), 313-314.

H. Shulman, Covering dimension and characteristic classes for foliations, Proc. Symp. Pure Math., Stanford 1976, Providence, Vol. 32, Part 2 (1978), 189-190.

H. Shulman and J. Stasheff, De Rham theory for BΓ, Springer Lecture Notes in Math. 652(1978), 62-74.

H. Shulman and D. Tischler, Leaf invariants for foliations and the van Est isomorphism, J. Diff. Geom. 11(1976), 535-546.

C. Siegel, Note on differential equations on the torus, Ann. of Math. 46(1945), 423-428.

J. Sikorav, Formes différentielles fermées sur le n-tore, Comment. Math. Helv. 57(1982), 79-106.

E. Silberstein, Multifoliations on $M^n \times S^2$ where M^n is a stably parallelizable manifold, Proc. London Math. Soc. 35(1977), 463-482.

M. Simonnet, Feuilletages et sous-fibrés intégrables, Esquis. Math., No. 28(1977), 1-65.

K. Sithanantham, On the cohomology of the Lie algebra of formal vectorfields preserving a flag, Ill. J. Math. 28(1984), 487-494.

J. Smith, Extending regular foliations, Ann. Inst. Fourier 19(1969), 155-168.

V. Solodov, On mappings of the circle into foliations, Geometrich. Metody v Zadachakh Analiza i Algebry, Yaroslavl' (1978), 100-107.

V. Solodov, A geometric proof of a theorem of Mather, Uspekhi Mat. Nauk. 34(1979), 243-244.

V. Solodov, Foliations on manifolds whose fundamental group has a nontrivial center, Uspekhi Mat. Nauk 36(1981), 229-230.

V. Solodov, Foliations on manifolds with a special fundamental group, Uspekhi Mat. Nauk. 36(1981), 225-226.

V. Solodov, Components of topological foliations, Mat. Sb. 119(1982), 340-354, 477.

V. Solodov, Homeomorphisms of the circle and foliations, Izv. Akad. Nauk. SSSR Ser. Mat. 48(1984), 599-631.

B. Solomon, On foliations of $\mathbb{R}^{n+1}$ by minimal hypersurfaces, Comment. Math. Helv. 61(1986), 67-83.

J. Sondow, When is a manifold a leaf of some foliation?, Bull. Amer. Math. Soc. 81(1975), 622-624.

D. Stasheff, Continuous cohomology of groups and classifying spaces, Bull. Amer. Math. Soc. 84(1978), 513-530.

P. Stefan, Accessible sets, orbits, and foliations with singularities, Proc. London Math. Soc. 29(1974), 699-713.

P. Stefan, Integrability of Systems of Vector Fields, J. London Math. Soc. 21(1980), 544-556.

S. Sternberg, Local C^∞ transformations of the real line, Duke Math. J. 24(1957), 97-102.

D. Sullivan, A new flow, Bull. Amer. Math. Soc. 82(1976), 331-332.

D. Sullivan, A counterexample to the periodic orbit conjecture, Inst. Hautes Etudes Sci. Publ. Math. 46(1976), 5-14.

D. Sullivan, Cycles for the dynamical study of foliated manifolds and complex manifolds, Invent. Math. 36(1976), 225-255.

D. Sullivan, Infinitesimal computations in topology, Inst. Hautes Etudes. Sci. Publ. Math. 47(1977), 269-331.

D. Sullivan, A foliation of geodesics is characterized by having no tangent homologies, J. Pure Appl. Algebra 13(1978), 101-104.

D. Sullivan, A homological characterization of foliations consisting of minimal surfaces, Comment. Math. Helv. 54(1979), 218-223.

D. Sullivan and R. Williams, Homology of attractors, Topology 15(1976), 259-262.

D. Sundararaman, Compact Hausdorff transversally holomorphic foliations, Springer Lecture Notes 950(1982), 360-372.

D. Sundararaman On holomorphic vector fields (maps) with singularities (fixed points) of Poincaré type, Coll. on dyn. systems, Guanajuato, 1983; Aportaciones Mat., Soc. Mat. Mexicana 1985, 127-151.

V. Surygin, Families of totally geodesic hypersurfaces in Weyl spaces, Trudy Geom. Sem. Kazan Univ. 10(1978), 140-147.

H. Sussmann, Orbits of vector fields and integrability of distributions, Trans. Amer. Math. Soc. 180(1973), 171-188.

T. Suwa, A theorem of versality for unfoldings of complex analytic foliation singularities, Invent. Math. 65(1981/82), 29-48.

T. Suwa, Residues of complex analytic foliation singularities, J. Math. Soc. Japan 36(1984), 37-45.

T. Suwa, Unfoldings of meromorphic functions, Math Ann. 262(1983), 215-224.

T. Suwa, Kupka-Reeb phenomena and universal unfoldings of certain foliation singularities, Osaka J. Math. 20(1983), 373-382.

T. Suwa, Unfoldings of foliations with multiform first integrals, Ann. Inst. Fourier 33(1983), 99-112.

T. Suwa, Unfoldings of complex analytic foliations with singularities, Japan J. Math. 9(1983), 181-206.

T. Suwa, Singularities of complex analytic foliations, Proc. Symp. Pure Math. 40, Part 2, 1983, 551-559.

H. Suzuki, Characteristic classes of foliated principal GL_r-bundle, Hokkaido Math. J. 4(1975), 159-168.

H. Suzuki, A property of characteristic class of an orbit foliation, London Math. Soc. Lect. Note, Ser. No. 26(1977), 190-203.

H. Suzuki, Construction of transverse projectable connections in some foliated bundles, Publ. Res. Inst. Math. Sci. 17((1981), 215-233.

H. Suzuki, Foliation preserving Lie group actions and characteristic classes, Proc. Amer. Math. Soc. 85(1982), 633-637.

S. Tabachnikov, Characteristic classes of homogeneous foliations, Uspekhi Mat. Nauk 39(1984), 189-190.

S. Tabachnikov, Characteristic classes of Grassman bundles, Funktsional. Anal. i Prilozen 19(1985), 83-84.

B. Tabak, A geometric characterization of harmonic diffeomorphisms between surfaces, Math. Ann. 270(1985), 147-157.

R. Takagi and S. Yorozu, Minimal foliations on Lie groups, Tôhoku Math. J. 36(1984), 541-554.

M. Takeuchi, On foliations with the structure group of automorphisms of a geometric structure, J. Math. Soc. Japan 32(1980), 119-152.

I. Tamura, Spinnable structures of differentiable manifolds, Proc. Japan. Acad. 48(1972), 293-296.

I. Tamura, Every odd dimensional sphere has a foliation of codimension one, Comment. Math. Helv. 47(1972), 164-170.

I. Tamura, Foliations of total spaces of sphere bundles over spheres, J. Math. Soc. Japan. 24(1972), 698-700.

I. Tamura, Foliations and spinnable structures of manifolds, Ann. Inst. Fourier 23(1973), 197-214.

I. Tamura, The Topology of Foliations, Iwanami Shoten Publ., Tokyo, (1976).

I. Tamura (Ed), Foliations, Proc. Symp. Univ. of Tokyo, July 1983.

I. Tamura and T. Mizutani, Null-cobordant codimension one foliation on S^{4n-1}, J. Fac. Sci. Univ. Tokyo, 24(1977), 93-96.

I. Tamura and A. Sato, On transverse foliations, Inst. Hautes Etudes Sci. Publ. Math. 54(1981), 205-235.

C. Tanasi, Sur les feuilletages mesurés arationnels, Cahiers Topol. Géom. Diff. Catégories 25(1984), 303-310.

C. Tanasi, Arational measured foliations, II, Rend. Circ. Mat. Palerms 34(1985), 300-309.

S. Tanno, A theorem on totally geodesic foliations and its applications, Tensor 24(1972), 116-122.

S. Tanno, Totally geodesic foliations with complete leaves, Hokkaido Math. J. 1(1972), 7-11.

D. Tanré, Groupes feuilletés, Esquis Math. No. 20(1973), 1-35.

C. Terng, Natural vector bundles and natural differential operators, Amer. J. Math. 100(1978), 775-828.

R. Thom, Généralisation de la théorie de Morse aux variétés feuilletées, Ann. Inst. Fourier 14(1964), 173-190.

R. Thom, On singularities of foliations. In: Manifolds Tokyo 1973, 171-173, Tokyo: Univ. of Tokyo, (1975).

R. Thom, Limit sets of leaves of foliations, Sûgaku 30(1978), 132-136.

W. Thurston, Non-cobordant foliations on S^3, Bull. Amer. Math. Soc. 78(1972), 511-514.

W. Thurston, Foliations of 3-manifolds which are circle bundles, Thesis, University of California, Berkeley (1972).

W. P. Thurston, Foliations and groups of diffeomorphisms, Bull. Amer. Math. Soc. 80(1974), 304-307.

W. Thurston, The theory of foliations of codimension greater than one, Comment. Math. Helv. 49(1974), 214-231.

W. Thurston, A generalization of the Reeb stability theorem, Topology 13(1974), 347-352.

W. Thurston, The theory of foliations of codimension greater than one, Proc. Symp. Pure Math., Stanford, Calif., 1973, Providence, vol. 27, Part 1 (1975), 320.

W. Thurston, A local construction of foliations for three-manifolds, Proc. Symp. Pure Math., Stanford, Calif., 1973, Providence, vol. 27, Part 1 (1975), 315-319.

W. Thurston, On the construction and classification of foliations, Proc. Int. Congr. Math., Vancouver, vol. 1, S. 1(1975), 547-549.

W. Thurston, Existence of codimension one foliations, Ann. of Math. 104(1977), 249-268.

W. Thurston, Hyperbolic structures on 3-manifolds, II: Surface groups and 3-manifolds which fiber over the circle, Ann. of Math., to appear.

M. Tibar, The graph of a Reeb foliation, Stud. Cerc. Mat. 36(1984), 262-266.

W. Ting, On nontrivial characteristic classes of contact foliations, Proc. Amer. Math. Soc. 75(1979), 131-138.

D. Tischler, On fibering certain foliated manifolds over S^2, Topology 9(1970), 153-154.

D. Tischler, Manifolds M^n of rank n - 1, Proc. Amer. Math. Soc. 94(1985), 158-160.

Ph. Tondeur, The mean curvature of Riemannian foliations, Colloque de géométrie symplectique et physique mathématique, Lyon 1986, to appear.

Ph. Tondeur and G. Toth, On transversal infinitesimal automorphisms for harmonic foliations, Geometriae Dedicata 24(1987), 229-236.

A. Torpe, K-theory for the leaf space of foliations by Reeb components, J. Funct. Anal. 61(1985), 15-71.

F. Torres Lopera, Grassmann structures and transversally Grassmann foliations, Proc. of the ninth Spanish-Portuguese Conf. on Math., Salamanca, 1982; Acta Salmanticensia Ciencias 46(1982), 549-552.

P. Trauber, The continuous cohomology of the Lie algebra of vector fields on a smooth manifold, Thesis, Princeton University (1973).

A. Treibergs, Entire spacelike hypersurfaces of constant mean curvature in Minkowski space, Invent. Math. 66(1982), 39-56.

F. Tricerri, Sulla geometria differenziale delle variete multifogliettate, Bolletino U.M.I. 16(1979), 76-84.

G. Tsagas, Some properties of closed 1-forms on a special riemannian manifold, Proc. Amer. Math. Soc. 81(1981), 104-106.

G. Tsagas, On the foliation of a Riemannian regular s-manifold, Tensor 36(1982), 150-154.

T. Tsuboi, Foliations with trivial $\mathcal{F}$-subgroups, Topology 18(1979), 223-233.

T. Tsuboi, On 2-cycles of BDiff(S^2) which are represented by foliated S^2-bundles over T^2, Ann. Inst. Fourier 31(1981), 1-59.

T. Tsuboi, Homology of diffeomorphism groups, and foliated structures (Japanese), Sûgaku 36(1984), 320-343.

T. Tsuboi, Γ_1-structures avec une seule feuille, Astérisque 116(1984), 222-234.

T. Tsuboi, Foliated cobordism classes of certain foliated S^2-bundles over surfaces, Topology 23(1984), 233-244.

N. Tsuchiya, Leaves of finite depth, Japan, J. Math. 6(1980), 343-364.

N. Tsuchiya, Leaves with nonexact polynomial growth, Tôhoku Math. J. 32(1980), 71-77.

N. Tsuchiya, Growth and depth of leaves, J. Fac. Sci. Univ. Tokyo, Sect. I.A. Math. 26(1979), 473-500.

N. Tsuchiya, The Nishimori decompositions of codimension-one foliations and the Godbillon-Vey classes, Tôhoku Math. J. 34(1982), 343-365.

I. Vaisman, Sur la cohomologie des variétés Riemanniennes feuilletées, C. R. Acad. Sci. Paris 268(1969), 720-723.

I. Vaisman, Sur la cohomologie des variétés analytiques complexes feuilletées, C. R. Acad. Sci. Paris 273(1971), 1067-1070.

I. Vaisman, Sur l'existence des operateurs différentiels feuilletés à symbole donné, C. R. Acad. Sci. Paris 276(1973), 1165-1168.

I. Vaisman, Cohomology and differential forms (1973), Dekker, New York.

I. Vaisman, Remarks about differential operators on foliate manifolds, An. Sti. Univ. Iasi 20(1974), 327-350.

I. Vaisman, From the geometry of Hermitian foliate manifolds, Bull. Math. Soc. Sci. Math. RSR 17(1973), 71-100.

I. Vaisman, On the differential geometry of the transverse bundle of a foliation, Rev. Roum. Math. Pures appl. 20(1975), 89-101.

I. Vaisman, A class of complex analytic foliate manifolds with rigid structure, J. Diff. Geom. 12(1977), 119-131.

I. Vaisman, Conformal foliations, Kodai Math. J. 2(1979), 26-37.

I. Vaisman, The Bott obstruction to the existence of nice polarizations, Mh. Math. 92(1981), 231-238.

I. Vaisman, A note on projective foliations, Publ. Sec. Mat. Univ. Autònoma Barcelona 27(1983), 109-128.

I. Vaisman, Obstructions to the existence of transverse volume elements of foliations, Dekker Lecture Notes in Pure and Appl. Math. 90(1984), 525-534.

T. Van Duc, Fibrés vectoriels feuilletés, Boll. Unione Mat. Ital. A15 (1978), 52-60.

T. Van Duc, Fibrés vectoriels feuilletés, Kodai Math. J. 1(1978), 205-212.

T. Van Duc, Connexions induites et classes caractéristiques, Atti Accad. Naz. Lincei Rend. Cl. Sci. Fis. Mat. Natur. 63(1977), 513-517.

T. Van Duc, Connexions induites et classes caractéristiques, Acta. Math. Vietnam. 3(1979), 23-27.

T. Van Duc, Feuilletage et distributions définies par une connexion, Rev. Roumaine Math. Pures Appl. 25(1980), 1019-1025.

T. Van Duc, Structure presque-transverse, J. Diff. Geom. 14(1979), 215-219.

T. Van Duc, Feuilletages transversalements tangents, Annali di Mat. 136(1984), 227-239.

J. Vanzura, A note on product structures and the Gelfand-Fuks cohomology, Colloq. Math. Soc. Trans. Bolyai 31(1982), 755-768.

P. Ver Eecke, Introduction à la théorie des variétés feuilletées, Amiens (1982).

P. Ver Eecke, Sur le groupe foundamental d'un feuilletage, C. R. Acad. Sci. Paris 300(1985), 55-58.

P. Ver Eecke, On a theorem of Charles Ehresmann concerning foliations, Cahiers Topologie Géom. Diff. 22(1981), 453-455.

P. Ver Eecke, Sur le groupe fondamental d'un feuilletage, Cahiers Topol. Géom. Diff. Catégoriques 25(1984), 381-428.

P. Ver Eecke, Sur le classifiant dn groupöide d'holonomie d'un feuilletage, C. R. Acad. Sci. Paris 300(1985), 639-642.

L. Verstraelen, Foliations of quasi-umbilical submanifolds, Simon Stevin 51(1977/78), 65-69.

J. Vey, Quelques constructions relatives aux Γ-structures, C. R. Acad. Sci. Paris 276(1973), 1151-1153.

E. Vidal, Sur les feuilletages réguliers et les problèmes qui s'y rapportent, Acta Cientifica Compostelana III (2) 69078(1966).

E. Vidal, Mesures défines sur les espaces des feuilles d'un feuilletage, Rend. Circ. Mat. Palermo 15(1966), 247-256.

E. Vidal, Sobre los sistemas diferenciales completamente integrables regulares, VI Reunión Mat. espanoles, Sevilla (1967), 73-75.

E. Vidal, Sobre algunos problemas en relacion con la medida en espaciios folados, In: I Coloquio Internacional de Geometria Diferencial, pp. 63-77, Santiago Universidad, (1964).

E. Vidal, Sur les variétés à structure de presque produit complexe avec métrique presque feuilletée C. R. Acad. Sci. Paris 273(1971), 1152-1155.

E. Vidal, Sur les connexions basiques transversales et projectables dans les variétés presque-produit complexes, C. R. Acad. Sci. Paris 277(1973), 461-463.

E. Vidal, Metricas casi-foliadas en variedades con estructura casi-producto compleja, Actas de las primeras jornalas mat. luso-espanolas (1973), 279-284.

E. Vidal, On regular foliations, Ann. Inst. Fourier 17(1967), 129-133.

E. Vidal and E. Vidal Costa, Special connections and almost foliated metrics, J. Diff. Geom. 8(1973), 297-304.

S. Vishik, On characteristic classes and singularities of foliations, Funkts. Anal. Ego Prilozhen. 6(1972), 71-72.

S. Vishik, Singularities of analytic foliations and characteristic classes, Funkts. Anal. Ego Prilozhen 7(1973), 1-15.

J. Viviente, Sur la géométrie transverse des feuilletages, Comun. I.C.M. Varsovia 1982.

J. Viviente, Una excursion por la teoria de foliaciones, Academia de Ciencias Exactas, Fisicas, Quimicas y Naturales de Zaragoza, 1984.

J. Viviente, Estructura cociente e invariantes básicos, Contribuciones Matematicas en honor del profesor D. F. Botella Raduan, Madrid 1986, 237-249.

E. Vogt, Stable foliations of 4-manifolds by closed surfaces. I. Local structure and free actions of finite cyclic and dihedral groups on surfaces, Invent. Math. 22(1973), 321-348.

E. Vogt, Foliations of codimension two with all leaves compact, Manuscr. Math. 18(1976), 187 212.

E. Vogt, A periodic flow with infinite Epstein hierarchy, Math. 22(1977), 403-412.

E. Vogt. Foliations of codimension 2 with all leaves compact on closed 3-, 4-, and 5-manifolds, Math. Z. 157(1977), 201-223.

E. Vogt, The first cohomology group of leaves and local stability of compact foliations, Manuscripta Math. 37(1982), 229-267.

E. Vogt, Comparison between the Kodaira-Spencer and Heitsch invariant associated to deformations of foliations, Resultate Math. 8(1985), 88-91.

A. Wadsley, Geodesic foliations by circles, J. Diff. Geom. 10(1975), 541-549.

E. Wagneur, A generalization of Novikov's theorem to foliations with isolated generic singularities, Topology and Its Appl., Proc. Conf. Mem. Univ. Newfoundland,k St. John's, Canada, 1973, New York (1975), 189-198.

E. Wagneur, Réduction des points singuliers des feuilletages à singularités non dégénérées de M^3, Can. Math. Bull. 19(1976), 221-230.

P. Walczak, On foliations with leaves satisfying some geometrical condition, Dissertationes Math. (Rozprawy Mat.) 226(1983), 47 pp.

P. Walczak, On minimal Riemannian foliations, Inst. Math., Pol. Acad. Sci., Warsaw 253(1981).

P. Walczak, Mean curvature functions for codimension one foliations with all leaves compact, Czech. Math. J. 34(1984), 146-162.

A. Walker, Connexions for parallel distributions in the large I, Quart. J. Math. Oxford 6(1955), 301-308.

A. Walker, Connexions for paralles distributions in the large II, Quart. J. Math. Oxford 9(1958), 221-231.

G. Wallet, Nullité de ℓ'invariant de Godbillon-Vey d'un tore, C. R. Acad. Sci. 283(1976), 821-823.

G. Wallet, Godbillon-Vey invariant and commuting diffeomorphisms, Differential Topology Varenna (1976), 151-159.

G. Wallet, Holonomy and vanishing cycle, Ann. Inst. Fourier 31(1981), 181-186.

G. Whiston, Cobordism through one-codimensional foliations, J. Diff. Geom. 11(1976), 475-478.

T. Willmore, Parallel distributions on manifolds, Proc. London Math. Soc. 6(1956), 191-204.

T. Willmore, Connexions for systems of parallel distributions, Quart. J. Math. Oxford 7(1956), 269-276.

T. Willmore, Systems of parallel distributions, J. London Math. Soc. 32(1957), 153-156.

T. Willmore, Connexions and foliated structures, Sitzungsber, Berlin Math. Ges., S. 1, S. A. (1969-1971), 14-15.

F. Wilson, Vector fields tangent to foliations, II, Handlebody foliations, J. Diff. Eq. 27(1978), 46-63.

H. Winkelnkemper, The graph of a foliation, Ann. Global Anal. Geom. 1 (1983), 51-75.

H. Winkelnkemper, The number of ends of the universal leaf of a Riemannian foliation, Diff. Geom. Maryland 1981/82, Birkhäuser Progr. Math. 32(1983), 247-254.

R. Wolak, On V-G-foliations, Rend. Circ. Mat. Palermo 1984, Suppl. No. 6, 329-341.

R. Wolak, On transverse structures of foliations, Rend. Circ. Mat. Palermo 1985, Suppl. No. 9, 227-243.

R. Wolak, Some remarks on V-G-foliations, Pitman Research Notes 131(1985), 276-289.

R. Wolak, On G-foliations, Ann. Polon. Math. 46(1985), 371-377.

S. Wong and S. Walter Wei, Bernstein conjecture in hyperbolic geometry, Seminar on minimal submanifolds, Ann. of Math. Studies 103(1983), 339-358.

J. Wood, Foliations on 3-manifolds, Doct. Diss. Berkeley Univ. Calif. (1968).

J. Wood, Foliations on 3-manifolds, Ann. of Math. 89(1970), 336-358.

J. Wood, Foliations of codimension one, Bull. Amer. Math. Soc. 76(1970), 1107-1111.

S. Yamagami, Cech cohomology of foliations and transverse measures, Proc. Japan Acad. Ser. A Math. Sci. 58(1982), 258-261.

K. Yamato, Qualitative theory of codimension-one foliations, Proc. Japan. Acad. 48(1972), 356-359.

K. Yamato, Qualitative theory of codimension-one foliations, Nagoya Math. J. 49(1973), 155-229.

K. Yamato, Examples of foliations with nontrivial exotic characteristic classes, Proc. Japan. Acad. 50(1974), 127-129.

K. Yamato, On exotic characteristic classes of conformal and projective foliations, Osaka J. Math. 16(1979), 589-604.

K. Yamato, Sur la classe caractéristique exotique de Lazarov-Pasternack en codimension 2, C. R. Acad. Sci. Paris 289(1979), 537-540.

K. Yamato, The Lazarov-Pasternak exotic characteristic class in codimension 2, II, Japan J. Math. 7(1981), 227-256.

K. Yano, Topological entropy of foliation preserving diffeomorphisms, Proc. Amer. Math. Soc. 85(1982), 293-296.

K. Yamato, Examples of foliations with nontrivial exotic characteristic classes, Osaka J. Math. 12(1975), 401-417.

S. Yorozu, Notes on square-integrable cohomology spaces on certain foliated manifolds, Trans. Amer. Math. Soc. 255(1979), 329-341.

S. Yorozu, Nonexistence of nonzero foliated harmonic forms on a compact foliated manifold, Ann. Sci. Kanazawa Univ. 18(1981), 11-17.

S. Yorozu, Behavior of geodesics in foliated manifolds with bundle-like metrics, J. Math. Soc. Japan 35(1983), 251-272.

S. Yorozu, The non-existence of Killing fields, Tôhoku Math. J. 36(1984), 99-105.

S. Yorozu, The second fundamental form of a foliation with parallel mean curvatures along the leaves, Ann. Sci. Kanazawa Univ. 22(1985), 1-10.

A. Zeghib, Feuilletages géodésiques des variétés localement symétriques et application, Thèse Dijon, 1985.

R. Zimmer, Ergodic theory, semisimple Lie groups, and foliations by manifolds of negative curvature, Inst. Hautes Études Sci. Publ. Math. 55(1982), 37-62.

R. Zimmer, On the Mostow rigidity theorem and measurable foliations by hyperbolic space, Israel J. Math. 43(1982), 281-290.

R. Zimmer, Curvature of leaves in amenable foliations, Amer. J. Math. 105(1983), 1011-1022.

R. Zimmer, Arithmeticity of holonomy groups of Lie foliations, preprint.

E. Zuzoma, A topological classification of foliations described by one-dimensional Pfaffian forms on an n-dimensional torus, Differential Equations 17(1981), 898-904.

E. Zuzoma, Singular Reeb foliations on an n-dimensional torus, Mat. Zametki 30(1981), 123-128, 155.

SUBJECT INDEX

adapted connection	47
associated transverse field	118
basic form	118
basic Laplacian	154
Bernstein theorem	98
Bott connection	47
bundle-like metric	51
characteristic form	65
closed manifold ($\equiv$ compact manifold without boundary)	10
comparison theorem	164
De Rham-Hodge decomposition	155
developing map	146
eiconal	113
exterior calculus	22
flat bundle	35, 37
flat connection	37
flow	132
foliation (definition)	24
force field	11
geodesible flow	134
G-foliation	49
Godbillon-Vey class	16
harmonic basic forms	164
harmonic foliation	68
holonomy	36, 38
holonomy invariant transversal volume	50, 72
holonomy invariant transversal measure	72

infinitesimal automorphism	117
integrability (≡ involutivity)	9,12
isometric flow	136
isoparametric function	116
isotopy	44
Killing vector field	51
Kronecker line	25
leaf	24
level hypersurfaces	104
Lie foliation	143
liquid crystal	5
magnetic monopole	88
Maurer-Cartan form	143
Maxwell's equation	89
mean curvature form	65
mean curvature vector field	65
Moser's technique	44
normal bundle of a foliation	26
projectable vector field	118
Reeb foliation	10
Riemannian foliation	51
Roussarie foliation	18
Rummler's formula	66
second fundamental form	62
spectral sequence of $\mathcal{F}$	120
symplectic foliation	125
tangent bundle of a foliation	26
tangential geometry	3,47
tangential Levi-Cività connection	49

taut flow	134
taut foliation	93
thermodynamical system	11
Tischler's theorem	41
totally geodesic foliation	58
transition function	25
transversal divergence	127
transversal geometry	3,47
transversal Levi-Cività connection ($\equiv$ unique metric torsion-free connection in Q)	53,54
transversal manifold	3,38
transversal homogeneous	147
transversal Killing field	129
transversally metric infinitesimal automorphism	129
transversally orientable	29,50
transversally oriented	8,51
transversally symmetric	164
trivial $(p + r)$-form	121
twisted duality	149,159
Weingarten map	62

INDEX OF NOTATIONS

$\check{B} \times_\Gamma F$	36
d	23
d_B	119
d_κ	153
$\operatorname{div}_B Y$	127
$F^r \Omega^m$	120
g_Q	51
$H_B \equiv H_B(\mathcal{F})$	119
Hess_f	104
$i(X)$	23
L	8, 26
L^*	27
$L^\perp$	47
Q	26
Q^*	20, 27
$V(\mathcal{F})$	118
$W(Z)$	62
Y	118
α	62
$\alpha^\perp$	63
$\langle \alpha, \beta \rangle_B$	149

$\|\alpha\|$	150
ΓL, $\Gamma(U,L)$	26
Γq^L	118
Δf	106
Δ_B	154
δ_B	152
$\theta(X)$	23
κ	65
$\kappa^\perp$	70
ν	50
τ	65
$\tau^\perp$	70
$\chi_\mathcal{F}$	65
$\chi(M)$	12
$\Omega^{\cdot}(M)$	22
$\Omega_B^{\cdot} \equiv \Omega_B^{\cdot}(\mathcal{F})$	119
$\overset{\circ}{\nabla}$	47
∇^M	48
∇	48
∇^L	59
∇f	104
g	143
$*$	149

LHS
  ($\equiv$ left hand side)

RHS
  ($\equiv$ right hand side)

ISBN 0-387-96707-9
ISBN 3-540-96707-9